THEORY OF SCIENCE AND TECHNOLOGY TRANSFER AND APPLICATIONS

Systems Evaluation, Prediction, and Decision-Making Series

Series Editor

Yi Lin, PhD

Professor of Systems Science & Economics
School of Economics and Management
Nanjing University of Aeronautics and Astronautics

Efficiency of Scientific and Technological Activities and Empirical Tests
Hecheng Wu, Nanjing University of Aeronautics & Astronautics
ISBN: 978-1-4200-8846-5

Grey Game Theory and Its Applications in Economic Decision-Making
Zhigeng Fang, Nanjing University of Aeronautics & Astronautics
ISBN: 978-1-4200-8739-0

Hybrid Rough Sets and Applications in Uncertain Decision-Making
Lirong Jian, Nanjing University of Aeronautics & Astronautics
ISBN: 978-1-4200-8748-2

Irregularities and Prediction of Major Disasters
Yi Lin, Nanjing University of Aeronautics and Astronautics
ISBN: 978-1-4200-8745-1

Optimization of Regional Industrial Structures and Applications
Yaoguo Dang, Nanjing University of Aeronautics & Astronautics
ISBN: 978-1-4200-8747-5

Systemic Yoyos: Some Impacts of the Second Dimension
Yi Lin, Nanjing University of Aeronautics and Astronautics
ISBN: 978-1-4200-8820-5

Theory and Approaches of Unascertained Group Decision-Making
Jianjun Zhu, Nanjing University of Aeronautics & Astronautics
ISBN: 978-1-4200-8750-5

Theory of Science and Technology Transfer and Applications
Sifeng Liu and Zhigeng Fang, Hongxing Shi and Benhai Guo,
Nanjing University of Aeronautics & Astronautics
ISBN: 978-1-4200-8741-3

THEORY OF SCIENCE AND TECHNOLOGY TRANSFER AND APPLICATIONS

SIFENG LIU · ZHIGENG FANG
HONGXING SHI · BENHAI GUO

CRC Press is an imprint of the
Taylor & Francis Group, an **informa** business
AN AUERBACH BOOK

MATLAB® is a trademark of The MathWorks, Inc. and is used with permission. The MathWorks does not warrant the accuracy of the text or exercises in this book. This book's use or discussion of MATLAB® software or related products does not constitute endorsement or sponsorship by The MathWorks of a particular pedagogical approach or particular use of the MATLAB® software.

CRC Press
Taylor & Francis Group
6000 Broken Sound Parkway NW, Suite 300
Boca Raton, FL 33487-2742

First issued in paperback 2019

© 2010 by Taylor & Francis Group, LLC
CRC Press is an imprint of Taylor & Francis Group, an Informa business

No claim to original U.S. Government works

ISBN-13: 978-1-4200-8741-3 (hbk)
ISBN-13: 978-0-367-38506-4 (pbk)

This book contains information obtained from authentic and highly regarded sources. Reasonable efforts have been made to publish reliable data and information, but the author and publisher cannot assume responsibility for the validity of all materials or the consequences of their use. The authors and publishers have attempted to trace the copyright holders of all material reproduced in this publication and apologize to copyright holders if permission to publish in this form has not been obtained. If any copyright material has not been acknowledged please write and let us know so we may rectify in any future reprint.

Except as permitted under U.S. Copyright Law, no part of this book may be reprinted, reproduced, transmitted, or utilized in any form by any electronic, mechanical, or other means, now known or hereafter invented, including photocopying, microfilming, and recording, or in any information storage or retrieval system, without written permission from the publishers.

For permission to photocopy or use material electronically from this work, please access www.copyright.com (http://www.copyright.com/) or contact the Copyright Clearance Center, Inc. (CCC), 222 Rosewood Drive, Danvers, MA 01923, 978-750-8400. CCC is a not-for-profit organization that provides licenses and registration for a variety of users. For organizations that have been granted a photocopy license by the CCC, a separate system of payment has been arranged.

Trademark Notice: Product or corporate names may be trademarks or registered trademarks, and are used only for identification and explanation without intent to infringe.

Library of Congress Cataloging-in-Publication Data

Theory of science and technology transfer and applications / Sifeng Liu ... [et al.].
p. cm. -- (Systems evaluation, prediction, and decision-making series)
Includes bibliographical references and index.
ISBN 978-1-4200-8741-3 (hardcover : alk. paper)
1. Technology transfer--China. 2. Science and state--China. 3. Technology and state--China. I. Liu, Sifeng.

T174.3.T483 2009
338.95106--dc22 2009030789

Visit the Taylor & Francis Web site at
http://www.taylorandfrancis.com

and the Auerbach Web site at
http://www.auerbach-publications.com

Contents

Preface

Technology, whose importance to domestic economic development has been widely recognized, has become the main strategy for competition across nations. With the rise in knowledge-based economies and the fast pace of technology development worldwide, there is a growing trend to develop high-end technologies, notably in information technology. Accompanied by economic globalization, the rising trend of globalization in technology enables the rational allocation and flow of the elements of technology without restrictions, allows the sharing of technological activities, and the space flow of technology more frequently. Technology has made a huge impact on economies and societies through technology transfer, which is regarded as an essential step for technology to have a social and a economic value.

Technology transfer has been given considerable importance in China as it is the key for improving core competence and is fundamental for the implementation and transfer of technological innovations to production. However, technology transfer has long been the weak link in establishing a national innovation system and is a great handicap for improving the self-innovation abilities of businesses because of the lack of proper mechanisms, regulations, and policies. To promote knowledge flow and technology transfer, it is important to fully utilize governments, colleges, scientific institutions, and businesses to explore and improve the effective mechanism of technology transfer.

Starting with the concept of technology transfer and its mechanism as the main theme, several issues, such as the measurement, the cost and benefit of technology transfer, the dynamics of the technical diffusion field, optimal allocation of technology transfer, and its game theory, are discussed in this book. Meanwhile, with some empirical studies based on the reality of China's technology transfer, some adventurous attempts and innovations have been made in both theoretical and practical aspects, which reveal the mechanisms, features, effects, and modes of technology transfer. All the studies involved provide constructive suggestions on implementing China's technology transfer in a reasonable way.

The nine chapters of this book are written by the following authors: Chapters 1 and 3 are written by Sifeng Liu, Chapters 4 and 8 by Zhigeng Fang, Chapters 2 and 7 by Hongxing Shi, Chapters 5 and 9 by Benhai Guo, and Chapter 6 by Lizhong Duan.

Jeffrey Forrest, Bingjun Li, Yaoguo Dang, Hecheng Wu, Lirong Jian, Ruilan Wang, Ying Wang, Chuanmin Mi, Jie Wu, Aiqing Ruan, Yanhui Chen, Yingying Ren, Xin Wu, Sandang Guo, Sha Li, Hui Zhou, Hongjiang Yue, Xiaogang Guo, Kun Hu, Chaoqing Yuan, Xiaohua Qin, Hanbin Kuang, Shawei He, Hongyu Hu, Yedong Wang, Qian Chen, Yong Liu, Xiaoyan Qiao, Yaping Li, and Yifan Zhang have taken part in related studies. Professor Sifeng Liu took charge of the draft summarization and final approval.

The research referred to in this book was supported by the National Natural Science Foundation of China; the Social Science Foundation of China, Jiangsu Province; the Soft Science Foundation of China, Jiangsu Province; Postdoctoral Programs Foundation of China, Jiangsu Province; Doctoral Programs Foundation of the Chinese National Educational Ministry; the Social Science Foundation of the Chinese National Educational Ministry; the Research Foundation for Excellent and Creative Teamwork in Science and Technology in Higher Institutions and distinguished professors of Jiangsu Province; and the Foundation for the Master's and Doctoral Programs in the Eleventh Five-Year Plan of Nanjing University of Aeronautics and Astronautics.

Any errors or omissions that may be pointed out by readers and experts in this field will be appreciated.

MATLAB® is a registered trademark of The Mathworks, Inc. For product information, please contact:

The Mathworks, Inc.
3 Apple Hill Drive
Natick, MA 01760-2098 USA
Tel: 508-647-7000
Fax: 508-647-7001
E-mail: info@mathworks.com
Web: www.mathworks.com

Acknowledgments

The research described in the book was supported by the National Natural Science Foundation of China (No. 70473037 and No. 70701017); the Key Project of Philosophic and Social Sciences of China (No. 08AJY024); the Key Project of Soft Science Foundation of China (2008GXS5D115); the Foundation for Doctoral Programs (200802870020); the Foundation for Humanities and Social Sciences of the Chinese National Ministry of Education (No. JA630039); the Key Project of the Soft Science Foundation of Jiangsu Province (No. BR2008081); and the Foundation for Humanities and Social Sciences of Jiangsu Province (No. 07EYA017). The authors would also like to acknowledge the partial support of the Science Fund for the Distinguished Professor of Nanjing University of Aeronautics and Astronautics (NUAA) and Jiangsu Province (No. 1009-316011), and the Science Fund for the Excellent and Creative Group in Science and Technology of NUAA and Jiangsu Province (No. Y0553-091). In the writing of this book, the authors consulted widely, referred to the research by many scholars, and were helped greatly by professor Jeffrey Forrest. We wish to thank them all.

Abstract

Starting with the concept of technology transfer and its mechanism as the main theme, several topics, such as the measurement, the cost and benefit of technology transfer, the dynamics of the technical diffusion field, the optimal allocation of technology transfer, and its game theory, are discussed in this book. Although there are some empirical studies based technology transfer in China, some innovations have been made in both theoretical and practical aspects, which reveal the mechanisms, features, effects, and modes of technology transfer. All of these studies provide constructive suggestions on implementing China's technology transfer in a reasonable way.

This book can be used as a textbook for postgraduates or senior undergraduate students specializing in economics and management and as a reference book for those who are involved in management, scientific research, and engineering technology.

Authors

Sifeng Liu received his bachelor's degree in mathematics from Henan University, Kaifeng, Henan, China, in 1981, his MS in economics, and his PhD in systems engineering from the Huazhong University of Science and Technology, Wuhan, Hubei, China, in 1986 and 1998, respectively. He has been to Slippery Rock University in Slippery Rock, Pennsylvania and Sydney University in Sydney, New South Wales, Australia as a visiting professor. At present, Professor Liu is the director of the Institute for Grey Systems Studies and the dean of the College of Economics and Management of Nanjing University of Aeronautics and Astronautics (NUAA); he is also a distinguished professor and guide for doctoral students in management science and systems engineering disciplines.

Dr. Liu's main research activities are in grey systems theory and regional technical innovation management. He has directed more than 50 projects at the national, provincial, and ministerial levels, has participated in international colloboration projects, and has published over 200 research papers and 16 books. Over the years, he has been awarded 18 provincial and national prizes for his outstanding achievements in scientific research and applications, has been appointed as a distinguished professor, and is a scholar and an expert who has made outstanding contributions. In 2002, Dr. Liu was recognized by the World Organization of Systems and Cybernetics.

Dr. Liu is a member of the committee of teachers of management science and engineering of the Ministry of Education, China. He also serves as an expert on soft science at the Ministry of Science and Technology, China, and at the Department of Management Science, National Natural Science Foundation of China (NSFC). Professor Liu currently serves as the chair of the Technical Committee (TC) of the IEEE SMC on Grey Systems; the president of the Grey Systems Society of China (GSSC); a vice president of the Chinese Society for Optimization, Overall Planning and Economic Mathematics (CSOOPEM); a vice president of the Beijing Chapter of the IEEE SMC; a vice president of the Econometrics and Management Science Society of Jiangsu Province (EMSSJS); and a vice president of the Systems Engineering Society of Jiangsu Province (SESJS). He serves as a member of the editorial board of over 10 professional journals, including *The Journal of Grey System* (Buckinghamshire, United Kingdom); *Scientific Inquiry* (Slippery Rock, Pennyslvania); *Journal of Grey System* (Taiwan, China); *Chinese Journal of Management Science*; *Systems Theory and Applications*; *Systems Science and Comprehensive Studies in Agriculture*; and the *Journal of Nanjing University of Aeronautics and Astronautics*.

Dr. Liu has won several accolades such as the "National Excellent Teacher" in 1995, "Excellent Expert of Henan Province" in 1998, "Excellent Science and Technology Staff in Jiangsu Province" in 2002, "Expert Enjoying Government's Special Allowance" in 2000, "National Expert with Prominent Contribution" in 1998, "National Advanced Individual for Returnee" and "Achievement Award for Returnee" in 2003, and "Outstanding Managerial Personnel of China" in 2005.

Zhigeng Fang received his master's degree in management science from the Xi'an Science and Engineering University in 1996 and his PhD in management science and engineering from Nanjing University of Aeronautics and Astronautics (NUAA) in 2006. Dr. Fang is currently working as a professor and a deputy director of the Institute for Grey Systems Studies, and as an assistant dean of the College of Economics and Management at the NUAA. He is the director of the Chinese Society of Optimization, Overall Planning and Economic Mathematics; the executive director of the Complex Systems Research Committee; the secretary-general of the Grey Systems Society of China; and the deputy director of the Jiangsu Post-Evaluation Research Center. He is also a member of Services Science Global and the IEEE Intelligent Transportation Systems Council. Dr. Fang's research interests include project management, post-evaluation, and grey systems. He has completed 26 academic and research projects for the country, military, provincial, and municipal departments, and has published over 70 research papers and six books. Over the years, Dr. Fang has been awarded eight provincial and national prizes for his outstanding achievements in scientific research and applications.

Hongxing Shi received his master's degree in management science from Nanjing University of Aeronautics and Astronautics (NUAA) in 1996. He is now a PhD candidate at the Institute for Grey Systems Studies at the NUAA. His main research activities are in grey systems theory, and the science and management of technology. In recent years, Shi has published two books and more than 20 papers. He participated in three projects at the national level and in five projects at the provincial level, and has been awarded the High-Quality Social Science Application Research Award of the Jiangsu Province.

Benhai Guo received his master's degree in management science from the Hefei University of Technology, Anhui Province. He is an associate professor at the Anhui University of Technology. Guo is now a PhD candidate at the College of Economics and Management, Nanjing University of Aeronautics and Astronautics.

His main research interest is in grey system theory, technological management, and energy economics. Guo has published three books, and over 20 papers in both national and international journals. He also participated in nearly 20 projects at the national, provincial, and ministerial levels including those of the National Natural Science Foundation and the National Social Science Foundation.

Lizhong Duan received his doctorate in management science and engineering from Tianjin University in 2001. He was engaged in postdoctoral research in systems engineering at the Nanjing University of Aeronautics and Astronautics from 2001 to 2003. He is now an associate professor at the School of Management, Beijing University of Chinese Medicine. Dr. Duan is a member of the standing committee of the Grey System Society of China and the Chinese Society of Optimization, Overall Planning and Economical Mathematics. His main research interests include technological diffusion and technological innovation management, and the application of grey system theory in medical diagnosis and management. He has published two books and more than 20 papers.

Chapter 1

Summary of Technology Transfer

1.1 Definition of Technology Transfer

1.1.1 Basic Meaning of Technology

People's understanding of technology is diverse; the definitions of technology are varied too.

The original meaning of technology was proficiency. In the proverb "Practice makes perfect," "perfect" meant technology. Now, it is understood that technology is a process or a process system. Diderot, a distinguished enlightened thinker, materialist philosopher, and education theorist in France in the eighteenth century, gave a concise definition of technology in the *Encyclopedia*: "The technology is varieties of tools and rules system that is collaborated for a common purpose." Some scholars also define technology on the basis of its purposes, components and functions, such as Gaynor (1996) who pointed out that technology can be described in different ways: first, that technology is the realization of resources into products or services; second, that technology includes knowledge and resources, which can help to reach established goals; third, that technology is an entity of science and engineering, which can be used in production processes and product designs, and also in the exploration to gain new knowledge. Li Ping (1999) considers technology as an effective means used by people to engage in various economic activities in spite of scarce resources, and the extension of technology includes products, processes, human resource, and organizations. Huang Jingbo (2005) holds that technology is a combination of knowledge, methods, skills, and special know-how that is used by humans in understanding and utilizing nature.

In essence, technology is a kind of systematic expertise associated with production processes of goods and services, and is a combination of the means, methods and skills created and developed by humans, to realize the needs of society. In terms of social productivity, the overall technical forces include technical skills, work experience, information knowledge, and equipments of solid tools, namely technical personnel, technical equipment, and technical information in the whole society. The other characteristics of technology include purpose, sociality, and pluralism.

Any new technology arises for a purpose, and the purpose of that technology runs through the entire process of technical activities. Hence, modern technology has strong utility and commercial features. The sociality of technology requires collaboration with the community and social support; it is also subject to a variety of social conditions. These social factors directly affect the success of technology and the development process. The pluralism of technology ensures that it can be expressed not only as tools for tangible equipment, machinery, entity material, and other hardware; but also as processes, methods, rules, and other knowledge software, as well as information and design drawings that are not material entities in themselves but material carriers of other manifestations.

1.1.2 Establishment and Evolution in the Definition of Technology

The descriptions of technology transfer can be summarized as follows:

Technology transfer is the flow of technology and is an important means of technology development.

Technology transfer is the transfer, proliferation, promotion, and transplantation of technical achievements in different countries, regions, sectors, industries, or enterprises.

Technology transfer is the flow of technology in different sectors, regions, and enterprises. Through technology transfer, technology combines with production processes to form new combinations and systems of technology. As a result productivity is enhanced, and economic benefit is improved continuously.

Technology transfer refers to organized delivery activities between supply and demand. In the process of technology transfer, the two sides are mutually constrained and interrelated. As a dynamic process, the realization of technology transfer is the result of the joint efforts of both supply and demand.

Technology transfer refers to the transfer of technology from production to application, making full use of the technology and realizing its value. Technology transfer includes the combination, transplantation, transmission, communication, and popularity of technology.

The international community discussed the issue of technology transfer for the first time in the first session of the United Nations Conference on Trade and Development in 1964. The meeting defined the input and output of technologies among countries as technology transfer. It is important to understand that technology transfer is not the physical transfer from point A to point B. It is also necessary to ascertain the systematization and complexity of transfer activities to make the connotation of technology transfer clear. As understood universally, technology transfer can be expressed as a certain type of technology-based diffusion process, which represents a certain technical level of knowledge. The United Nations' definition, in "International technology transfer in the draft code of conduct," refers to it as the transfer of system knowledge on the manufacture of a product, producing methods or providing service, but does not include the sale of goods or simply rent. Two representative views are more enlightening: the first is the definition made by Rose Bloom of Harvard University. In his view, technology is acquired, developed, and utilized through a path which is entirely different from its origin, and this process of technical change is technology transfer. The definition stresses the point that we must emphasize the adaptability of technology and environment in this shift, and not simply the move from one place to another. The other view was put forward by the American scholar Sipei, S.A. Based on the organized thought of human behavior promoted by anthropologist Harrington, technology transfer is organized work to achieve the goal and to make the necessary technical information move reasonably. He limits it to the planned and rational flow between the government and enterprises, and emphasizes its orderliness and regulations.

In the late 1970s, the concept of technology transfer came to China. The time that the Chinese were introduced to this theory can be traced back to as early as 1978. Tang Yunbin (1978) quoted the definition made by H. Brooks of Harvard University. Brooks holds technology transfer is knowledge developed in a certain group or agency, but this kind of knowledge is realized in matters of other groups or agencies. There are two types of transfers in his opinion; one is vertical transfer, which refers to knowledge transferred from general and common areas to more specialized fields, often from the basic new scientific knowledge to the field of application technology, namely the broader technology transfer. The other one is horizontal transfer, it refers to technology transfer from one application to another application, in the narrow sense of technology transfer. This horizontal transfer is divided into three categories: the first is achieved through trade, namely buyers introduce advanced technology through the purchase of advanced products.The second is the pure or original meaning of technology transfer, which is the technical trade of licensing trading. The buyers introduce technology to self-production, using the technology itself for trading. This is the so-called turnkey package deal. The contents of transfer include production technology and the corresponding equipment, plant, as well as staff training, and production preparation.

In 1982, the *Outlook* magazine no. 8 published an article "What is technology transfer?" to explain the term. In the interpretation, technology transfer was defined as the transfer of the results of science and technology, information, the transfer of ability, transplantation, import, promotion and popularization, and so on. For example, industrial production will put raw materials, technology, equipment, products, drawings, technical programs, design, as well as some theoretical research results and ideas for exchange in areas, or departments. Scientific technology transfer can happen in many forms. There are two general types and five channels. Of the so-called two types, one is inner transfer, and the other is transfer to production applications. The five channels are from the laboratory to production (including basic research, applied research, application of research and development to production), from military to civilian application, from the advanced region, sector, or industry to backward regions, sectors, industries, from the urban to rural areas, from domestic to foreign countries.

From the 1980s, many Chinese scholars began to study the issue of technology transfer. The most representative of their work is the gradient theory of technology transfer promoted by Yulong Xia and Zhongxiu He. In March 1982, Yulong Xia of Shanghai Institute of Science published an article named "Gradient theory and regional economics" in the magazine *Research and Suggestion* and promoted "gradient theory." In the same year, Zhongxiu He of the Tianjin Science and Technology Commission submitted a paper "On the gradient transmit of technology" at the World Assembly of Sociology. It was the first time that the "Domestic technology transfer to promote the law of gradient" was promoted. They pointed out that there is a natural gradient distribution in the economic and technical development in China—coastal, central, and remote areas—due to the economic and social imbalance. The mainland and a number of remote areas are rich in resources. But owing to historical reasons, insufficient funds, and slow development, technological development is poor, resulting in quite a few areas still in the traditional and backward economy. The central region is in the middle level whereas the coastal areas are equipped with "advanced technology"; the strength of their economy is obvious. Domestic technology should be adopted by the technical services, the transfer of outcome, compensation trade, joint ventures, and joint companies to achieve the transfer gradient of technology, namely, transferring "advanced technology" to "intermediate" and "traditional" technology. Later, the theory of evolution became the guiding ideology of China's macroregional development. Some academics have suggested a gradient transfer theory with a different view. Yuan Gangming (1997) pointed out that China's western region can introduce and develop advanced technology on their own and do not have to accept the technology delivered by the main coastal areas. In the western region, there have been a large number of successful cases of introducing and developing advanced technology to go beyond gradient sequencing and get faster development. The gradient policy favors the developed regions and delays the development of underdeveloped areas, which will only increase regional disparity. Kai Liang and Lianshui Li (2005) put forward the counter-gradient theory and pointed out that less developed areas should develop in leaps and bounds, taking advantage of the income tax preferential policies extended to technology transfer in the Chinese Corporate sector to assess the effect of analysis using cost–profit analysis, charts and draws analysis, and other methods.

In the late 1990s, the academic community reached a consensus: the time and space spread of technology was defined as technology diffusion, the transfer from the laboratory to the production unit was referred to as technology transformation, technology owners granted the right of application to other people; all of which is collectively referred to, in the broad sense, as technology transfer. This book is in favor of such a consensus.

The connotation of technology transfer is still in the development stage while the study of technology transfer theory has been developed further. With worldwide competition in science and technology, technology transfer research has a wide range of areas for research and an equally wide scope for increase in the content.

1.1.3 Discriminate Correlative Definition of Technology Transfer

1.1.3.1 Technology Innovation

The concept of technology innovation was firstly proposed by economist, Joseph Schumpeter. Schumpeter used the word "innovation" for the first time in his book *Theory of Economic Development* (German), which was published in 1911 followed by an English edition in 1934. The book reads: Innovation is the economic system, which took place in a serious deviation from equilibrium. It cannot make a new equilibrium in the balance on the basis of the old through the gradual adjustment to achieve. In 1928, Schumpeter published a new article "Instability of capitalism" in the *Journal of Economics*, which discussed innovation for the first time. In his view, innovation is a productive resource for innovative applications, with nongradual characteristics; innovation needs a large amount of venture capital before benefits can be reaped, the resistance to success stems from the unprecedented and uncertain nature of the innovative activities. In 1939, Schumpeter systematically described the theory of innovation in his book *Business Cycle*, where innovation is described as the introduction of a new economic system in the production function. It is mainly generated by the role of entrepreneurs. Innovation can be divided into technology innovation and nontechnology innovation. An important characteristic of his innovation theory is that he looks at it as the reason that capitalist society appeared a significantly nonbalanced economic cycle. In 1951, Solomon wrote an article titled "Capitalism in the process of innovation—on the theory of Schumpeter" in the *Economics Quarterly*. Solomon thought the significance of technology innovation is that it is a major source of economic change consisting of two parts—conceptualization and development work. It laid the foundation for the definition of future technology innovation. In 1953, anthropologist H.G. Barnett wrote a book *Innovation—The Basis of Cultural Change*. In his book, he analyzed innovation in sociology and psychology, and proposed that innovation is essentially systematic attention to the process, with special emphasis on the innovation of a process with a nonprogressive nature. In 1954, W.R. Maclaurim published the article "Invention and innovation to the order and its relationship with economic growth" in the *Economics Quarterly*. In his opinion, innovation emerges when invention is introduced into commercial applications in the form of a new or improved product or process. Moreover, he pointed out that invention and innovation are both completed at different stages by ordinary individuals or institutions, indicating that invention and innovation overlap between these two stages, but they have different characteristics. In 1962, J.L. Enos gave a complete definition of technology innovation for the first time in the "Oil-refining industry of invention and innovation," pointing out that innovation is the successful outcome of a series of activities, which include the search for innovation, the implementation of the funding, the establishment of organizations, employment of workers, the development of the market, etc. In 1963, a corporation named Arthur. D. Little separated the concepts of invention, innovation, and diffusion in an article titled "American industrial technology and innovative form and issue" for the United States National Science Foundation. In their opinion, technology is taking the lead in developing and refining inventions of practical value on the product or process ideas. Technology innovation is the invention for a commercial application. The proliferation of technology innovation is widespread. In 1974, C. Freeman in his book on "Economics of industrial innovation," defined technology innovation as new products, new processes, and new systems or the beginning of a new device from the laboratory to its commercial success in application and the whole of the activity process. In 1977, E. Mansfield published his book *New Industrial Technology in Production and Application*, in which he pointed out that innovation is the whole process of development of a new product from the time of the exploratory work until the new product is available for sale. At this point, the basic concept of technology innovation got a commitment, and formed a relatively consistent. In 1985, R. Moss of the Bell Labs Moss in the United States gave a "definition of technology innovation," published in *Project Management Journal*, pointing out that in their collection of about 350 articles on technology innovation literature, 75 percent showed striking consistency. This is now the definition of technology innovation—the process beginning from the idea of a new concept to the success in a meaningful way of the practical application of the idea for nontechnical phenomenon. Economists generally believe that productivity growth and a corresponding increase in the per capita income depends on the continuous process of technical change. This process, to a large extent, is reflected in the success of the new infrastructure development, production, and distribution. Wang Yingluo and Jia Liqun summed up all the past versions and defined technology innovation as follows: it includes new products, materials, technology, or other systems based on the direct use of natural and technical knowledge, as well as the application, development, design, drafting of the product specification, manufacturing production prototype, and preproduction processes.

1.1.3.2 Technology Diffusion

Technology diffusion is a topic widely explored by economists, and is closely linked with technology transfer, technology spillover, and many other related concepts. Technology diffusion is the simply the description of the movement of technology from one place to another, or from one user to another. In general, a technological innovation is limited in its economic and social impact on productivity improvement, and only when the new technology has been integrated into the production process will it maximize the potential economic benefits, to promote evolution of the economic system and gentrification.

Some scholars emphasize that technology diffusion is a simple acquisition not only for production technology but also to build the technical capacity of import. From the point of view of Balin Sen, the engineering and design ability of the local region are more important than merely acquiring knowledge and increasing production, as the region should have the capacity for technical change. In fact, the technology diffusion process is a learning process, which is an activity of independent continuous innovation on the basis of imitation. Learning reduces costs; the learning curve exists not only in individual learning, but also in the team's cooperation, organizations, and industries (Ping, 1999).

Some scholars also stress that technology diffusion is a process of selection. Metcalfe believes that technology diffusion includes the choice of enterprise at various levels of technology, as well as the selection of customers in enterprises. Because of these interactions of the selection process, the outcome of technology has increased dissemination in the market, so technology innovation is a step-by-step process to achieve technology diffusion.

Some scholars focus on the characteristics of technology diffusion and lay stress on the new process of imitation and the reinvention. They think dissemination, infiltration, and crossover of time and space constitute the essence of technology diffusion. This process is the movement of output and input through a variety of carriers in different countries, regions, industries, and enterprises. It is also primarily an invention for commercial use, in the process of continual re-innovation and constantly in increasing demand. Schumpeter also believed that the spread of technology innovation is essentially an act of imitation. Many companies, in pursuit of excess profits will join the ranks of imitators, because it is a wonderful role model, increases efficiency substantially, and reduces cost and a small number of enterprises, while imitating the process, will attempt a technological innovation to produce new products resulting in the diffusion process. Schumpeter saw the large-scale "copy" as technology diffusion. Only by spreading widely can any technology provide economic and social benefits.

Some scholars also define technology diffusion from a technical knowledge of the properties; it is essentially a process of knowledge exchange, a source of innovation, a potential source for current and potential users of the innovative coding knowledge and the implicit knowledge.

Rogers, who is a representative of the Communication Theory (Mansfield, 1971; Rogers and Scamell, 1990) believes that the proliferation of technology is the process of diffusion in the social community in a certain period of time through a channel in the system. The essence of the diffusion process is that the original message or new ideas are spread from one individual to another (Rogers, 1995). J.S. Colemen holds the view that technology diffusion is the process flow of information; it must be carried out in the middle of a bridge or a two-step flow. Chinese scholars Jiaji Fu and

Qingrui Xu share the same point of view; they believe that the diffusion process is the core of the bandwagon effect, namely the usage by potential adopters of the technology depends on the decision of the consumer. Mansfield, who proposed the model of imitation based on learning to imitate, pointed out that the technology diffusion process is a process of imitation—a proactive learning. When the copy contains progressive innovation, it becomes a high-level study. Therefore, he adopted the "infectious disease" model. P. David and S. Davies have put forward the stimulate learning—response mechanism theory. They suggest that adopters follow a "stimulus–response" mechanism in the process of technology diffusion. When stimulation of the potential adopters meets the "critical level", potential adopters respond to the stimulus with innovation. As a result of continuous diffusion, learning by doing leads to the reduction of marginal cost, and the "critical level" reduces, and in turn promotes future technology diffusion.

1.1.3.3 Technology Spillover

Technology spillover refers to the technology that companies promote locally through nonvoluntary diffusion of technology as productivity cannot increase with local technology alone. In the process of localization, it is a kind of external economic performance. Many research scholars work in the area of spillover effects. Since the 1990s, the latest theory and research about spillover are mainly in the following five areas:

1. The main figures of firm theory on the premise of spillover are Klibanoff, Morduch, Lee, Boisot, Poyago, and Theotoky. Klibanoff and Morduch (1995) examined technology spillover between companies. In the competitive model, the external economic and technical spillover led to an increase in inefficient firms, without a positive correlation with the external efficiency. But they found that with small spillover conditions, cooperation is of no real use, but in the large overflow, only through cooperation can the efficiency of the economy be improved. Lee (1995) found that manufacturers' research in and development of their technology and their contacts with the outside technology have become more active in the context of new technologies, compared to the small firms. Boisot (1995) discussed the effect of the neoclassical learning and Schumpeter learning on manufacturers' skills, utilizing cultural space as an analytical tool. Poyago and Theotoky (1995) studied the oligarchs wherever there was a joint venture of the best balance and scale. They found that the market would not be able to function but for the cooperation between manufacturers to provide adequate stimulation for the overflow of information in a simple model of oligopoly; so the joint R&D companies are usually smaller than the size of a balanced level of the best.
2. In the analysis of the spillover in game theory, Kapur and Ziss are the main figures. Capper (1995) studied the uncertainty of the manufacturer and

learned acts spillover by building a game model. Ziss (1994) built a two-stage game of a bilateral oligopoly model, comparing the noncooperative model with the joint venture (R&D phase of the collusion), price (production stage of collusion), and combination (R&D and production stage of the comprehensive collusion) of collusion. And then he assessed the condition in a variety of ways to improve the outcome of collusion. When the overflow is large enough, the outcome level in all three forms of collusion is higher than in the noncooperative manner; it is highest in the combined manner. The price is lowest in the case below.

3. Gugler, Dunning, Hagedoorn, and Duysters support the spillover analysis in a strategic alliance. Dunning and Gugler (1993) observe that a strategic alliance is to create, maintain, and improve technical advantages and include the vendor's innovative activities and regional allocation of complementary forms of organization; the spillover of the international technology of strategic alliance varies with each industry. The R&D integrated complex is part of the creation of technology, the proliferation of organizations, and organizational complexity of the network; strategic alliance can be integrated into the proliferation activity in the international study of mainstream economics with the theory of manufacture. Hagedoom (1995) studied the noncore technology strategy of the 1980s on cooperation between companies, and revealed the union's basic strategy for technology trends. Duysters and Hagedoom (1996) found on an analysis of the companies' R&D activities, innovation, and production strategies, and the international trend in technical cooperation that, even in global industries such as IT, the internationalization of innovation was still low.
4. Parente, Colombo, and Mosconi support the spillover effects and "learning by doing" theory. Parente (1994) studied the technology diffusion, learning by doing, and the relationship between economic growths. He chose specific manufacturers to study the time to absorb technology by the learning-by-doing model, by observing a variety of techniques that the manufacturers learnt by doing, through the acquisition of knowledge and know-how to improve the available technology. He also proved that the technology companies absorbed decision-making and output growth techniques in capital markets depending on their effectiveness. Colombo and Mosconi (1995) analyzed the compounded technology diffusion compatibility in the early years and the cumulative effect of learning later. Learning by doing is an effective tool of technology diffusion in the path of technical experience.
5. Mckendrick, Keluomubo, Buzzacchi, and Mariotti support the spillover analysis of organization technology. Their focus is on the organization technology of banks. Mckendrick (1995) found that the source of imitation for banking organizations technology is mostly nonmarket intermediaries. This source with expertise in various manufactures will, they expect, diversify in the near future. Buzzacchi, Keluomubo, and Mariotti (1995) believe that most of the

economists engaged in diffusion and innovation of technology and research take the manufacturing process only as an example and ignore the fact that the proportion of the services industry is rising continuously in developed countries. Hence, they discussed the diffusion of banking technology in Italy.

1.1.3.4 Inherent Relationship of Correlated Concept

1.1.3.4.1 Technology Transfer and Technology Diffusion

According to the earlier analysis, technology diffusion and technology transfer are interrelated, but there are obvious differences between the concepts. It is generally believed that the definition of technology diffusion is broader than that of technology transfer and that the applications of the two are also different. Technology transfer is mainly a means of the subjective purpose of economic activities. Both the sides involved in technology transfer have a clear purpose. Technology diffusion includes both the actual technology transfer and the incidental technology improvement, with more emphasis on the latter. Technology transfer involves the international community in general for the movement of technology, whereas technology diffusion is limited to the scope of technology in the domestic context. In some specific cases, technology diffusion has also been seen as technology transfer, but only when technology diffusion and technology transfer are directed at new technology or its application.

It is clear that technology diffusion and technology transfer have common links as well as clear distinctions.

The links between the two are as follows:

First, technology diffusion and technology transfer require movement of technology through certain channels in different geographical areas or between mobile technologies as a prerequisite for technology innovation, and its essence is the ability to transfer technology. When the movement is across borders, the movement becomes international technology diffusion and international technology transfer.

Second, the channels of diffusion and transfer of technology are generally identical; both include deliberate transfer of technology, and need-based "unintended" dissemination of technology.

Third, technology transfer and technology diffusion are complex processes; in addition to technology being the main link between the networks, they are both affected by technical, economic, social, and cultural concepts.

The introduction of the transfer to or diffusion in a region has the importer of the technology as the main participant; its ability to absorb the technology directly determines the outcome of the diffusion or transfer of technology.

Fourth, the complete process of technology diffusion and technology transfer comprise the process of being constantly used, copied, and emulated and reinnovated.

The differences between the two are as follows:

First, there is only one acceptor in the transfer of technology in general, and the target is clear. But there are several acceptors in technology diffusion, mainly the potential adopters.

Second, the transfer ends when the acceptors master the technology transferred, whereas diffusion ends when all the potential acceptors adopt the technology. Thus, more emphasis is placed on the latitude of time.

Third, technology diffusion is the process of spread and divergence, that lays stress on the concept of time and space and the external effects of technology. Technology diffusion and technology spillover and even the use of the two alternatives, are consequently linked. Technology transfer stresses the process of movement to a location, often ignoring the impact of external technology. In fact, the processes of technology transfer happen at the same time as the spillover of technology.

1.1.3.4.2 Technology Transfer and Innovation

Technology innovation and technology transfer areas are two complementary processes. Technology transfer cannot be separated from technology innovation and technology innovation is critical to technology progress as a whole. They form the fundamental mechanism in the progress of technology, the improvement of economics, and in socioeconomic development; they comprise the entire course of technology invention (the emergence of new technologies), technology development (new technology), and technology diffusion (the application of new technologies to promote). Technology innovation is a basic prerequisite and the main source for technology transfer. Beginning with the technical change in development, the two processes of technology innovation and diffusion run parallel to each other and promote each other. Technology innovation drives the process of technology transfer from the content as the source of the transfer. Technology transfer is the driving force behind technology innovation, and it not only provides technical resources but also energizes the act of innovation.

Technology diffusion makes the innovation more complete and effective. In some specific cases, technology diffusion has also been viewed as technology transfer, but only when technology diffusion and technology transfer are both directed at new technology or the application of new technology. However, the targets of transfer often refer to the existing technology and not new ones. This is precisely the reason why technology innovation and transfer of technology are considered as complementary processes.

1.1.3.4.3 Technology Innovation and Technology Spillover

To some extent, technology spillover inhibits the innovation enthusiasm of main enterprises with negative effects, which becomes especially apparent in the case of a large overflow of technology. Enterprises pursue high profits and low risks as the main theme of market competition; they will not carry through technology innovation contrary to this principle. With limited capacity in the market, the cost has become an important part of profit. The enterprises attempt to reduce the total cost of technology innovation in their own way, but the complexity of the task requires high R&D investment, and it is the complexity of technology and the high investment that determine the characteristics of high-risk. But technology has overtaken companies and made low-cost access to others' research results possible, thereby avoiding costly investment in research and development and reducing risk at the same time; besides, the result of technology innovation of enterprises with low-cost competitors has been to promote the level of competition, virtually reducing their competitive advantage, which is independent innovation with no interest in the results. Because all the decision-making businesses are affected, the ultimate result is that enterprises are willing to "wait," and reap the profits. The external effect of this technology spillover ends in market failure, and this cannot inspire enterprises to innovation.

It goes without saying that the negative effect of the above analysis is only from the point of view of costs and profits; the main consideration is the impact of technology diffusion on independent innovation. In fact, there are many factors that inspire enterprises' innovation. In real economic activity, many companies have growing enthusiasm for technology innovation, and innovation activities do not stall because of negative effects, the pressure of competition and patent protection system playing a key role. Patent protection system defines the ownership of technology outcome from the point of view of property rights, and those who imitate will be charged patent fees to make up for the spillover effects of independent innovation and the loss of business suffered to safeguard the interests of the innovator. But the patent system has defects, such as elimination of the monopoly of knowledge. The transaction costs are inevitable due to transfer among enterprises. Hence, when there is a more effective system or model, the technology spillover external effect is internalized.

1.2 Manner and Effect Factors of Technology Transfer

1.2.1 Basic Form of Technology Transfer—Technology Flow

Technology flow is the basic form of technology transfer. In theory, technology flow is associated with talent flow, material flow, capital flow, and information flow. The academia has a long history of research on technology diffusion and innovation and there are two schools of study of the dynamic mechanism and the channels of flow. For example, Chinese scholars in economic geography, have come up with a regional development point of the axis theory, gradient and antigradient theory with a broad spread of technology transfer, regional economic and technical development, and foreign direct investment (FDI) in theoretical research. Though these works continue to be macroscopic views, some of them can also be used for technical theory.

1.2.1.1 Components of Technology Flow

Technology transfer is the essence of technology flows; mainly three types of knowledge flow. Accordingly, technology flow also has three categories:

The first category is the physical flow of knowledge in terms of the products, parts and components, equipment, and manufacturers.

The second one is the invisible flow of knowledge in terms of know-how, patents, and other information, including technical data, documents, standards, technical manuals, service contracts, and maintenance manuals.

The third is the macro- and microflow of information in national, regional, business organizations, and individuals, because this knowledge can not only be clearly written in the form of text, but can also be operated in practice to understand and master.

Although there is no clear stipulation of the third knowledge in the transaction of flow, it is clear that it contributes to the flow knowledge in both the first and second categories. In fact, technology management of, production management of and marketing skills for the flow of knowledge are very difficult, as it depends on experience, learning, and innovation of the acceptors in practice.

1.2.1.2 Features of Technology Flow

Technology flow refers to the flow process of technology capacity, which covers technology innovation, technology diffusion, and technology transfer. Specifically, the technology flow has the following five features.

1. Systemic technology flow involves the technology, sender, channel, and receiver.
2. Initiative technology flow refers to capacity of technology, rather than the flow itself, including the ability to use the technology and innovate.
3. Bidirection technology flow is different from the concept of transfer and diffusion; it is not a one-way flow from the supplier of technology to the acceptors, but a two-way process with feedback, output, and other complex processes.
4. Interchangeable technology flow is the process of flow, where the source of technology could become the acceptor, and the technology acceptor can also become the source of technology. There is no definite boundary.
5. Dynamic technology flow is an endless dynamic process. Technology acceptors spread technology after accepting it, which develops into a spiraling dynamic process with no end.

1.2.1.3 Channels of Technology Flow

The technology flow channel is an integral part of the technology flow and the medium to connect the supplier and the acceptor. The complexity of the technology itself, the supply-side conditions (of the region), the level of technology, development strategies, and other factors contribute to create a wide range of technology flow channels.

Different levels of technology have different channels. Primary production's technology transfers are mostly turnkey projects, concessions, and management contracts. It is common for labor-intensive export industries to contract and import machinery and equipment, outsource foreign technology and licensing, and these are generally through channels to introduce the technology to regions, which are backward or have been using standardized technology. High-tech industries' flow channel of technically oriented technology alliance helps to reduce costs, risk diversification, and harmonize competition.

Different times give rise to different channels of technology flow. In developed countries, mergers and acquisitions were important channels for technology flow in the 1980s. In the 1990s, due to the integration of multimedia technology, technology alliance became the mainstream technology flow. Developing countries are still dealing in the purchase of equipment, FDI, and other main types.

The obligations of supply and demand and the ability to acquire technology are different in different channels, for example in the channel of trade in goods, machinery and equipment imports, FDI, technology alliances, mergers and other major flows, the technical capabilities and the degree of restraint are ordered from lower to higher ranks.

The success of technology flow is not determined by the kind of channel; the psychological and geographical approaches are critical success factors.

1.2.2 Ways of Technology Transfer

Technology transfer takes place between different countries and regions, enterprises, colleges and universities, research institutes, and others, where the enterprise is the most important subject. Technology transfer happens in a variety of specific ways, which can be roughly divided into two categories: internal and external technology transfer. Considering transnational corporations as an example for technology transfer, internal transfer is mainly through FDI, and external technology transfer happens in many ways, such as the sale of technology, licensing transactions, and technical assistance.

1.2.2.1 FDI

FDI refers to the direct capital investment in factories and mines of a country or a region of the production output by another country or region for the direct management of the factories and mines. FDI is made in four major forms. First, it holds wholly owned enterprises, of the capital that is generally entirely owned by investors of a country or region, foreign investment shares 100 percent; second, foreign ownership buys the stock and achieves a certain proportion of some or all of their control; third, investment and joint venture organized in the host country; fourth, investors reinvest their profits. As the technology is the premise of FDI of transnational corporations, the FDI in general combines technology transfer; in fact the FDI has become a major technology transfer method.

1.2.2.2 Sale of Technology

The sale of technology refers to the technology that multinational corporations sell as a separate production technology. This happens when the technology in the local market has no value, or the corporation is not prepared to reuse the technology in the future, and the technology is likely to become obsolete if it does not capitalize on the technology. To obtain the profits as soon as possible, transnational corporations will sell this technology in the market. Selling directly is profitable to both sides.

1.2.2.3 License Transaction

License transaction refers to sale by transnational corporations of patented technology, know-how, the right to use trademarks, product manufacturing, and marketing rights to other overseas subsidiaries or enterprises by signing the license contract and allowing the other party to use it. License transaction is the most common way for transnational companies to engage in technology transfer. License transaction can get the technology owned by others quickly, win time and save cost. It also helps to speed up the introduction of updated technology and improve the country's industrial structure.

1.2.2.4 Coproduction and Cooperative Research

Coproduction refers to the agreement between transnational corporations and host countries or regional enterprises, by which they produce jointly and come up with a reasonable division of labor in accordance with the production and management strengths. Coproduction technology can be provided by the transnational companies or they can research and design jointly. In this way, the technology transfer can take many concrete but flexible forms to ensure that the interests of both sides are realized.

Cooperative research happens when the two sides engage in design and research together and complete a certain project making use of the advantages of the host country or regional enterprise. In this period, the two sides can relate to each other's experience and technology. The two sides share the fruits of cooperative research and hold the patent right and copyright together.

1.2.2.5 Technology Assistance

This is a movement based on the complexity of the technology and the ability of the demand side. It is a more flexible way of technology transfer. In this way, the obstacles raised by the technology transfer can be conquered and the right to use the technology can be ensured for the recipient. Usually, technology assistance includes personnel training, technical advisory services, management consulting services, and marketing and business services.

1.2.2.6 Turnkey

A turnkey project is a visual comparison of technology transfer carried out by a whole set of engineering contracts. Technology acceptors are committed to the supplier for the contracted project, such as factory or plant. Technology suppliers are in charge of all the technology and project management of the project, from the design to the equipment and its installation until the test drive qualifies. Ultimately, they hand over the plant or factory, ready to start work at any time. As a result, it is actually a comprehensive international economic cooperation.

1.2.3 Influencing Factors of Technology Transfer

1.2.3.1 Law and Policy Factors

Technology is a kind of knowledge product; it needs to be protected by the legal system, particularly in the field of intellectual property rights. The degree of protection for intellectual property rights varies in different countries in the cross-border business environment. In some countries, a large number of counterfeit products infringe the international technology transfer and reduce the profit of export. Many companies in these countries can rarely get rid of the infringement problem; the protection of intellectual property rights of technology transfer in the host country or region is an important factor. In China, for example, there is a large gap in the level of protection of intellectual property rights between the market economy developed countries and China; on the whole, people's awareness of intellectual property rights are weak in China. Statistics shows that chemical and pharmaceutical industrial imitation rate is as high as 97 percent; intellectual property disputes occur frequently.

The policy system plays a decisive role in the effectiveness of technology transfer. It is difficult for technology transfer to succeed in a country or region, where there are frequent government interventions and many restrictions on foreign-funded enterprises. Policies of technology-importing countries or regions will have a direct impact on technology transfer in scientific and technical content and quality. For example, in the past the Chinese auto industry's policy was too strict for foreign enterprises to invest; many transnational companies shied away from investing. As a result, the introduction of China's first foreign car, Santana, monopolized the Chinese market. This imported car, the Volkswagen Santana, represented the technology of the 1970s. When placed in China, Volkswagen had condemned it. Since then, Santana was upgraded four times in Germany and in fact the life cycle of every generation of products is only four years. But the Santana in China is almost the same as the original 20 years ago! Santana is one of the examples of the condition that each upgrade should be approved by the local government.

1.2.3.2 Market Factors

Market factors are the fundamental factors of technology transfer. The effect of market competition and market size on technology transfer is particularly obvious. If the opening level of the home country or region is low, only a small number of transnational companies operating a monopoly in an industry, will be able to maintain the technical advantage of their monopoly status to gain more profits. So the industry slows down the speed of technical progress and hampers the technical development.

Taking the automotive industry as an example, in the past, because of China's auto import barriers and the market access restrictions, in the Chinese market Santana encountered a rival only in Peugeot. And soon after the latter's withdrawal from China, the lack of market competition seriously slowed the upgrading of the Santana. This is another reason a number of world-class automotive companies entered the Chinese market with China's reform, opening-up, and in-depth expansion, resulting in a fundamental change in competition. Competition among companies is the essence of technical strength. To gain a foothold in the market, enterprises have to update technology to speed up the pace of technology transfer.

The size of the market determines the scale of production. If the market capacity is large, economy grows steadily, and residents' purchasing power is strong in a country or a region, the growing space of technology-importing countries or regions will be great. Accordingly, these areas' attraction of investment and technology transfer will be stronger.

1.2.3.3 Technology Basis

A nation or region's technology basis is important for technology transfer. The availability of human resources, the knowledge level, the development of productive forces, and the technology level will have a real impact on the transfer. The countries and regions with a great technology base and high skill levels will be capable of exporting technology. The conditions of technology-importing countries or regions restrict their ability to accept new technologies. If other conditions remain constant, the country with high-performance technology and good technical basis is more likely to promote the country's technology transfer. Taking Motorola entering the Chinese market as an example, its first arrival in the Chinese market did not bring in the most advanced technology, which is one of the reasons why China's technical knowledge base is still weak. With the development in the economy and the enhanced technology capabilities, Motorola increased its investment in China and set up 18 R&D centers. The advanced technology moved to China gradually.

1.2.3.4 Infrastructure Status

Infrastructure includes transportation, canals, ports, bridges, telecommunications, electricity, water and urban water supply and drainage, gas, electricity, and other facilities. These are substantial engineering facilities to provide public service for production and the residents. It is the common material base for production, management, work and life, and guarantees that the main facilities in a city operate normally. Moreover, it is not only an important condition for material production, but also an important condition for reproduction of labor. As the exporter or recipient of technology, the infrastructure construction must be taken into consideration in the implementation of technology transfer.

1.3 Main Features of Technology Transfer

1.3.1 Characteristics of Contemporary International Technology Transfer

With the vigorous development of the new technology revolution and with world economic globalization deepening, international technology transfer develops rapidly in a variety of different forms. Economically backward countries are trying to develop their own economy through technology transfer. With the rapid advancement in technology transfer today, developed countries do not dare to ignore the transfer to maintain their leading position. The contemporary international technology transfer has shown the following characteristics.

1.3.1.1 Structure Upgrade of International Transfer Technology

With the fierce competition in technology and talents, the structure of international investment and technology transfer upgrade have steadied, whereas the proportion of knowledge-intensive style and software implementation has increased. In particular, the use of license trade to transfer technology with a higher degree of monopoly is increasing. This trend goes hand in hand with international direct investment.

The internalization of technology transfer is a feature of international transfer. To obtain high profits and strengthen the monopoly and protection of technology, transnational corporations who are the major sources of international technology transfer allocate the technology resources within the company to ensure they do not lose the technology, operation, and management advantages. As the control of technology strengthens, the mobility of capital and the global operation of transnational corporations increase. Direct investment has, therefore, become an important means of technology transfer.

Technology can flow independently, detached from FDI. This is reflected in the payment for the relevant intellectual property rights and professional services and the development of the relationship of nonsubsidiary strategic partners' alliance. The National and foreign companies conclude a direct technology transfer agreement. Among the peer transnational companies in operation with the capability and resource of technology, the technology strategic alliance, which aims to exploit new techniques, controls the new international standards and maintains the market competitiveness, came into being. It is also being developed as a new way for technology transfer.

1.3.1.2 Multipolarization of International Technology Transfer Supply

Technology innovation is a source of technology transfer. The continuous invention of modern science and technology and the shortened life cycle of new products and new technology help promote the transfer of technology. From the 1970s, an international network of technology innovation came into being. It continues to play an increasingly important role in technology revolution and innovations worldwide to multipolarize the technical supply of technology transfer. There are many manifestations of technology supply. From the point of view of form, it is common for transnational companies to achieve international technology transfer through strengthening international M&A (Mergers and Acquisitions). Some transnational companies set up overseas research and development institutions and take advantage of the host country's human resources, technology, facilities, education, research, and other resources, and develop suitable techniques for the host country in the economic development and trade-related development needs. Since the 1990s, the Asia-Pacific region has gradually become the emerging market in international technology transfer. The Asia-Pacific region's economic growth rate is higher than the rest of the world; international trade and investment activity flourish here. This becomes the important basis in the development and upgrade of FDI, trade and high-tech products, technology trade, and other transfer activities. In addition to passively acquiring direct investment from developed countries for new techniques, some newly industrialized countries take the initiative to invest in developed countries to get the necessary new techniques.

1.3.1.3 Development of Technology Transfer from Monodirection to Bidirection

In general, technology transfer has been a one-way transfer from developed countries to less developed countries, advanced regions to laggard regions, central departments to the external sectors, all of these showing the gradient. This gradient technology transfer not only considers the relative distance of the location, but also considers the economic and cultural differences and the natural environment. Therefore, developed countries have long been exporters of technology products, and developing countries the importers. This has formed a basic pattern with technology trade among the developed countries occupying the absolute position in international technology transfer. With the multipolarization of the International Center for technology innovation and the diversification of technology transfer supply, the flow of international technology transfer between the countries or regions or industries shows bidirection. On one hand, bidirection technology transfer among developed countries involves investment, technology trade, exchange, and technical cooperation; on the other hand, mutual technology transfer develops step by step between developed and developing countries. Bidirection transfer, complementary advantages, and storage with each other have become an inevitable trend in today's world technology transfer.

1.3.1.4 Transfer Technology, an Important Dynamic Move to Promote Economic Development

Along with the launch of the international technology transfer, the pace of technical progress accelerates even as labor productivity rises in many countries or regions'. Technical progress promotes industrial division of labor, which helps a number of industries improve their productivity and enhance their competitiveness. Consequently, it might change a country's comparative advantage and change the pattern of international competition. Technology transfer will cause changes in industrial structure and affect the industrial structure in both production and demand, either directly or indirectly. On the producers' side, the new technique created by the transfer will not only create new products and new industry, but also improve the labor productivity and financial profitability in existing industries. In the meantime, some backward, unprofitable productions or industries may die out. On the demand side, new production levels and living needs will be created continuously so as to meet the needs of these new industries; some out-of-date production and living needs will become extinct and the corresponding industries will be eliminated. At present, high-tech products have become the world's leading force in the development of trade. Modern technology trade, which deals with high-tech products as the object of expansion; international trade in technology service takes technology services and information transfer as content and develops fast. All the above shows that the functions of technology transfer to promote economic development have become increasingly significant.

In addition, international technology transfer also shows other new trends, such as the faster growth of transfer, expanding scale, broadening area, and gradient transfer. Transnational companies have become the main force of international technology transfer using their huge financial resources and strong technical force.

Internationalization of science and technology is common. International cooperation and technology exchanges strengthen and the protection of international technology transfer is enhanced. Governments' intervention in technology transfer is greater in degree than the general commodity trade.

1.3.2 Features of Technology Transfer in China in Different Periods

The compartmentalization of technology transfer development in China is varied in the academic field. Rongping Kang (1994) argued that, from the foundation of China, the history of the import of technology can be divided into three periods: the "Soviet Union mode" period (1950–1979), the transition period (1980–1989), and the new period (1990–). Jianguo Li (1997) partitioned it into three periods, the first period mainly imitation of the Soviet Union's mode, the second from the reform to the present, and the third from the present to the future. Linhai Wu and Huagui Zhu also divided the process into three steps, with the first period 1950–1978, second 1978–1991, third from 1992 to the new market-oriented economy.

From the available papers and materials, it is clear that researchers basically agree on these three periods theory. It is necessary to consider both the national and international circumstances in a specific period to compartmentalize a nation's development of technology transfer. On the basis of this consideration, this book divides it into four periods.

1.3.2.1 Period of Imitation (1949–1959)

In this period, advanced technology was mainly imported from the Soviet Union and other socialist countries in east Europe. International technology transfer dominated this period; later the imported techniques were promoted in China and modified to suit the practical domestic situation. During the first five year plan, around the beginning of the foundation of PRC (People's Republic of China), China's industries were in a mess and had no advanced techniques. Therefore, China imported massive equipments and the corresponding techniques to pave the road for the industrialization of the new China. Based on the statistics from the national project department, between 1950 and 1959 China signed for over 700 projects, 450 of which were imported and cost 3.7 billion dollars in total, which is roughly about 50 percent. During the first five year plan (1952–1957) 156 techniques were imported, which were also the keys projects in the process of industrialization of China such as coal, electric power, oil, metallurgy, chemical engineering, electric machine, aviation, automobile, light industry, textile, and military industry. With the import of those techniques, the pace of China's industrialization was hastened significantly; every section achieved its own magnificent and expected goals, and the gross industrial output value increased by 18 percent.

The features of this period are:

First, the techniques were only from the Soviet Union and some east European socialist countries.
Second, the object of the technology transportation was to help China build its own heavy industrial systems.
Third, the transportation was confined to buying the equipments and techniques as a complete set, and merely copying the industrial systems and standards in countries like the Soviet Union.
Fourth, the main body of the transportation was confined to the country.
Fifth, the transportation of technique was combined with the cultivation of talents in techniques and economic management.

1.3.2.2 The Fumbling Period (1959–1978)

In the late 1950s, due to the tension between China and the Soviet Union, the import of techniques from those countries became stagnant; the sequential three year long natural disaster brought China's economy to a very critical state. During this period the self dependence policy basically dominated, and the resistance to imported techniques increased significantly. Especially during the three-line construction period, an avalanche of new techniques and talents moved to the middle-east, promoting the diffusion of technique in the middle-east part of China. During the late 1960s and early 1970s, China gradually established diplomatic relations with European capitalist countries and Japan, and the technology transfer was carried out in nongovernmental individual exchanges, and many key techniques and equipments related to oil, chemical engineering, chemical fiber, metallurgy, mining, electronic, nicety apparatus, and textile mechanism were imported from France, Britain, and Japan and the import structure emphasized the importance of the light industry. As the Sino-US relationship began to thaw and the negotiation started, the technique exchange between China and the United States was given a boost.

The features of this period were:

First, the circumstances were very complicated, and the zigzag road of technique development, as well as the impact of the culture revolution, slowed the development. The understanding of the technique imported and international trade stayed at the entry level.
Second, technology transfer shifted from the domestic to the international only gradually, and China began to provide technique-aid to some Asian and African countries.
Third, with the development of technology transfer globally, China changed its technique mode from the earlier one of entire equipment import to the present one of key equipment import only, and began to enlarge the technique import channel by technique admission and technology service.

Fourth, the purpose of technique import was not only for building the heavy industry systems and producing homemade products by referring to previous industrial systems and standards (basically from Soviet Union), but also for enhancing the light industry.

Fifth, the main body of technology transfer was not confined to nations; the technology exchange among individuals was also enhanced.

1.3.2.3 Stage of Internalization of Development (1978–1992)

In the late 1970s, two important events had a significant impact on technology transfer in China—the establishment of diplomatic relations between China and the United States and China's economic reform and the opening-up of its economy. In more than a decade from the late 1970s to early 1990s, China imported critical economic construction, chemical, mechanical and electrical, and other aspects of the production from many other countries. This played a great role in promoting China's economic take-off. During this period, China's academic community started research based on international technology transfer, and technology transfer turned from international to domestic technology transfer issues. The access to a number of theoretical achievements helped to guide the practice of technology transfer at various levels. The Chinese government strengthened the development of combination. They believed that technology transfer must combine with independent innovation and international technology transfer must combine with domestic technology transfer. They used the gradient theory of technology transfer theory, and developed science and technology in China in a phased and conditioned manner, and diffused mature technology in domestic regions. All these actions effectively made the natural potential energy brought in by technology transfer into dynamic movement for economic development for the realization of the "three-step" strategy. At the same time, Chinese technology transfer policies and regulations improved gradually, and the technology market development standardized gradually. In October of 1980, the State Council promulgated the "Regulations on the development and protection of the Provisional Regulations of the socialist competition." For the first time transfer of technology was clearly defined. In 1980, State Science and Technology, the annual national work conference on science and technology drafted a report "on China Development of science and technology policy report on the outline." In 1983, the State Science and Technology Commission issued "to strengthen technology transfer and technical services," which marks the beginning of the technology transfer market.

The external influences of the political storm in 1989, as well as China's own economic restructuring caused a low ebb in the introduction of technology again; the developed countries blockaded China's economic technology, so China had to rely on independent innovation and technology transfer to develop domestic high-tech. As a result of this blockade, the computer, telecommunications, and home appliance industry in China got a period of respite for digesting the introduced excess techniques and for innovation. Lenovo, Changhong, Skyworth, and a large number of domestic enterprises were strengthened.

The features of this stage of the technology transfer include:

First, the theoretical research in technology transfer was strengthened. The transition from the importance of international technology transfer to research on domestic technology transfer issues was done. The theory began to bear results.

Second, the scale of companies and enterprises, which introduced technology help to expand them further, to develop from the formal instruction operation to the guidance operation and commission. The examining and approving authority of technology import project loosened its hold. The degree of autonomy improved by the introduction of technology.

Third, the manner of introduction was more flexible; the introduction of single set or key equipment in the past was fundamentally changed and replaced by more typical methods, such as technology licensing, technical services, technical advice, coproduction, compensation trade, equipment leasing, joint ventures, franchising, and complete sets of equipment and the introduction of key equipment, and so on. It expanded the channels for technology.

Fourth, technology tended to diversify sources of funding, government loans, special foreign exchange, domestic commercial loans, and loans from international financial institutions, foreign exporters of technology export credits, and local self-funded enterprises played a more important role in this.

Fifth, many enterprises have increased their own R&D efforts to improve the efficient use of technology.

1.3.2.4 Opening-Round Stage (1992–)

In 1992, China's reform of the economy system met another watershed, the Communist Party of China 14th National Congress put forward the goal of a socialist market and economic system, and China's economic system once again achieved a breakthrough. At the same time, the international situation has been improving, and international economic and technical exchanges redeveloped. Theorists began a multilevel, multiangle study of domestic and international technology transfer. Technology transfer policy and the building of the legal and economic environment for it continue to push forward. China's technology import has a heightened degree of openness more than ever before; the arrangements for technology transfer are increasingly rich, and the situation is even more diverse. On one hand, advanced science and the techniques introduced from the international community are digested and absorbed in proper areas; on the other hand, domestic scientists and technicians are encouraged to develop advanced technology and help industrialization and transfer and diffuse it to other parts of the country or abroad. This force of domestic technology is in a low state. Through the output of technology to foreign countries, the corresponding profits can be increased.

This stage of technology transfer is characterized by:

First, technology transfer theory has become more sophisticated, and actively guides the practice of technology transfer.
Second, technology transfer expands the scale of activities and brings about the development of Chinese-led multinational companies.
Third, enterprises have become the main body of technology transfer, and research and development activities are initiated in theory and practice.
Fourth, domestic capital market perfects itself increasingly, which provides abundant sources of funding.
Fifth, in China, the enterprises which are determined to compete in the international market are facing new challenges and opportunities. For example, Haier, Lenovo, Huawei, TCL, and many other enterprises have increased their own R&D efforts and have begun to use the rules of the market for technology mergers and capital acquisitions, in-kind M&A, and to accelerate the growth of the business scale and to start to export the technology.

Chapter 2

Technology Transfer and Its Mechanism, Characteristics, Effects, and Models

Technology transfer obeys the economic law, which is its underlying mechanism, to some extent, rather than the disorder and irregularity of free movement. In this chapter, we study the mechanism of technology transfer in terms of its direction, dynamics, vectors, and other aspects, analyze its characteristics, and consider the space division model, ultimately revealing the law of technology transfer.

2.1 Mechanism of Technology Transfer

Technology transfer refers to the horizontal transfer of technical elements and the regional implication prompted by the uneven distribution of technology resources, technology potential difference, and other factors. Its essence is the reconfiguration of some useful technical elements to suit different demands. Such a transfer is not a disorderly and irregular movement; it obeys some economic law, as shown in Figure 2.1, revealing the inherent mechanism of technology transfer.

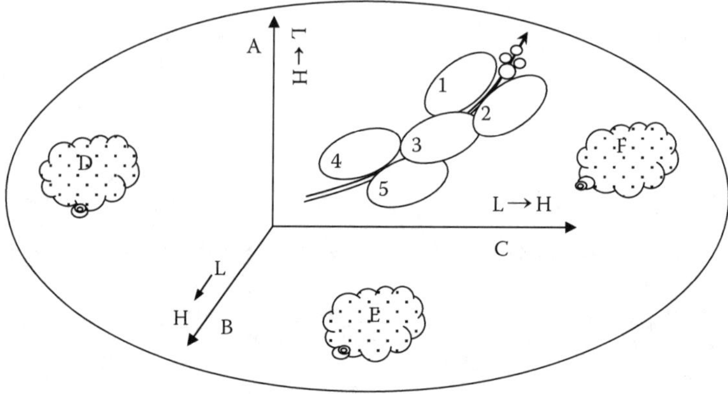

Figure 2.1 The schematic diagram for the mechanism of technology transfer. A, technical level; B, economic energy; C, benefit return; D, policy environment; E, living environment; F, investment environment; 1, information; 2, capital; 3, management; 4, material resources; 5, talent; L, low; H, high.

2.1.1 Local Potential Difference Determines the Direction of Technology Transfer

Owing to the uneven distribution of technology resources and the difference in the structure, scale, and level of long-term development, different regions show considerable differences in their stock of material resources, human resources, information resources, management resources, and other technical elements. This leads to "technology potential difference" in different areas and industries. Technology potential difference being a vector with directivity, technology transfer also has directivity, as technical elements only flow from regions at a higher technical level to one at a lower level (or from a region abundant in technical elements to a deficient region).

Technology is always easy to transfer between regions at almost the same technical levels. The greater the difference between the two regions at the technical level, the greater the resistance there will be to the transfer, and the lesser the possibility that the different technology will combine with the new productivity. Therefore, technology transfer and technical level have a nonlinear relationship.

2.1.2 Determination of Feature in Technology Transfer Elements by Margin of Interest

Laborers combine the means of labor with the subject of labor in the process of working, making the factors of matter "live," generating the energy, which can transform natural substances, to produce materials and realize productivity. There can be no productivity if there is only "matter" and no labor. So, the use of technical elements, whatever their properties and forms, is dominated and determined by the labor of human beings. Clearly humans control the transfer of technical elements directly or indirectly.

With the ongoing development of market economy, awareness of the economy is growing stronger. "Input–output benefit" has been the guideline in the transfer of technical elements. When all the conditions that influence technology transfer are the same, technical elements will flow to regions which have higher output benefit. The more the margin of interest, the more the capacity (or amount) of technical elements transferred. The margin of interest has been the driving force of technology transfer.

From the point of view of marginal cost and benefit, if the marginal benefit is more than the marginal cost, the profits of technology transfer will increase continuously, whereas if the technology transfer's marginal benefit is equal to the marginal cost, then the profits of technology transfer reach a maximum.

2.1.3 Technology Elements Are the Carriers of Technology Transfer

Technology transfer is the process where technical elements in some region flow to another region, to combine rationally with the productivity factors in the other region and improve technology and productivity. Material resources, human resources, information management, and other factors constitute technology transfer; they are the carriers of technology. In addition, the concrete form of technology in the process of the transfer reflects the properties of the technology transferred.

These technical elements form dynamical networks, transmit the technical ability, and serve as an arterial system for technology by means of cooperation (interregional trade, interregional coordination, interregional investment, etc.), and operation (scatter, concentration, insert, etc.), crossing the barrier of space and the boundaries of industry.

2.1.4 Principle of Energy Increase in Technology Transfer

On one hand, technical elements flow from a region at a higher technical level to a region at a lower level, leading to increase in productivity in the latter region and resulting in an increase in the technical elements transferred. On the other hand the technical elements are always attached to some concrete matter (talent, machine, etc.) in the transfer. The impact of the worker's processing and reprocessing will often change the chemical nature and efficiency quality of the output. Besides, the outflow of the technical elements (for example, material resources, information, etc.) will not lead to a lower output in the resource region but only increase the output in the receiving region. Therefore, technology transfer, rather than complying with the

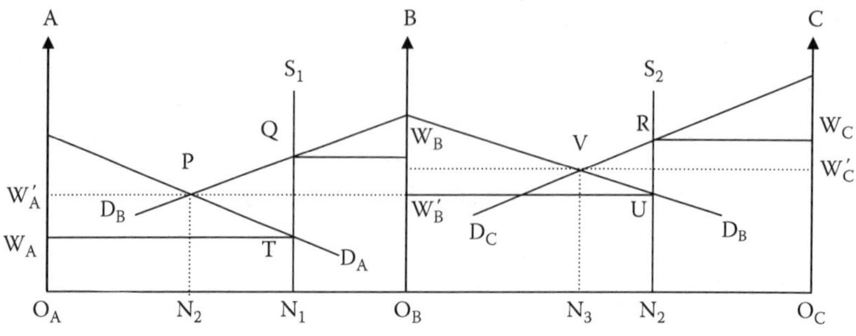

Figure 2.2 The principle of energy increase in technology transfer.

"law of conservation of energy" will actually increase the value of the community and promote social development, as shown in Figure 2.2.

In this figure, the axes O_A, O_B, and O_C represent the quantity of technical transfer, axes A, B, and C represent the yield as a result of the technology transfer that happened in the three regions, D_A, D_B, and D_C represent the demand curves for certain types of technology, and S_1, S_2 represent the supply curves (short-term supply is a vertical line due to the scarcity or short supply of technology). Affected by the difference in yields between A and B ($W_A < W_B$), when the technology flow is from region A to region B, the technical elements are reduced in region A and technical supply is reduced from the initial N_1 to N_2, which reduces the region's output Q_1, indicated by the area N_1TPN_2. As for region B, with the supply of technology increasing from N_2 to N_1, the effective size of technology increases leading to an increase in its output Q_2 indicated by the area N_1QPN_2.

It is clear from Figure 2.2 that technology transfer increased the social total output to Q_3 indicated by the area TQP, and that $Q_3 = Q_2 - Q_1$. By the same token, affected by the difference in yield between B and C ($W'_B < W_C$), when the technology flow is from region B to region C, the technology is reduced in region B, technical supply is reduced from the initial N_2 to N_3, which also reduces its output to Q_4 indicated by the area N_2UVN_3. For region C, the inflow of the technology increases from N_3 to N_2, and leads to an increase in its output Q_5 indicated by the area N_2RVN_3. The social total output increases to Q_6 indicated by the area URV, and $Q_6 = Q_5 - Q_4$. It is a one-way loss of technology for region A, which does not receive compensation; therefore, technology transfer has had a negative social effect on region A. It is a one-way flow of technology for region C; therefore, technology transfer has had a positive social effect in absorbing technology and increasing profit. For region B, there has been a two-way transfer of technology, both in and out. When $N_1 - N_2 > N_2 - N_3$, $Q_2 > Q_4$ implying that the loss of technology has been compensated in time with output increasing and hence technology transfer has had a positive social effect. On the other hand, when $N_1 - N_2 < N_2 - N_3$, $Q_2 < Q_4$, implying that the loss of technology has not been adequately compensated with output reducing and hence technology transfer has had a negative social effect. From the perspective of society as a whole, technology transfer increases the total output by Q_3+Q_6. This shows that technology transfer helps optimize the allocation of production factors, promotes the restructuring of the productive forces, promotes the rationalization of the economic structure, and affects the all-round development of socially productive forces.

2.1.5 Technology Transfer Has a Two-Way Network Trend

As countries around the world continue to open up by trading across borders and regions, the two-way transfer and cross-growth of technology have been enhanced. In different time periods, technical levels are different between any two regions. In the same period, not only do two regions with different industries have different levels of productivity, but the same interindustry productivity levels also vary. As a result, technology, rather than always flowing from one area to another, forms a dynamic two-way transfer at different times both with different industries and industries with different factors of production in the same area, as shown in Figure 2.3.

Western Europe in the 1950s and Japan in the 1960s relied mainly on the United States for transfer of technology. After the 1970s, however, the transfer of technology was from Western Europe and Japan to the United States. At present there is cross-transfer of technology between these countries and it is difficult to quantify the flow clearly. In China, advanced technology in the developed eastern coastal areas flow to the West; the production rich resources of the West continue to flow to the East, forming a two-way transfer.

Owing to a variety of factors, conditions, and constraints, technology transfer is a complex multifactor process. When a combination of factors that influence the transfer is greater in a particular region than in other regions, the technology will

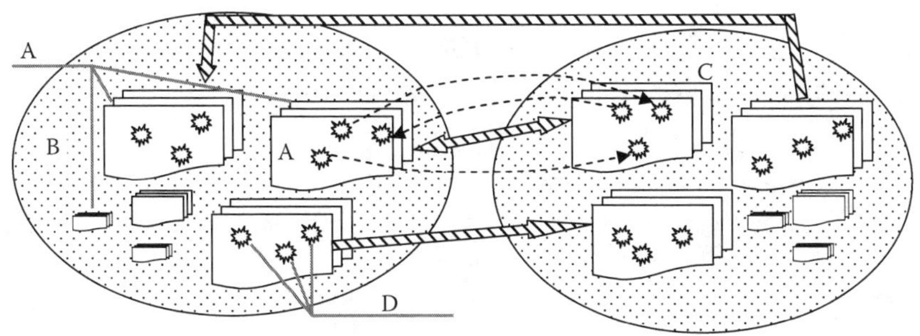

Figure 2.3 Two-way transfer of technology. A, industry; B, region 1; C, region 2; D, production elements. *Note*: The oval map stands for the region, the flowchart for the industry, the star for production elements, and the arrow for the direction of transfer.

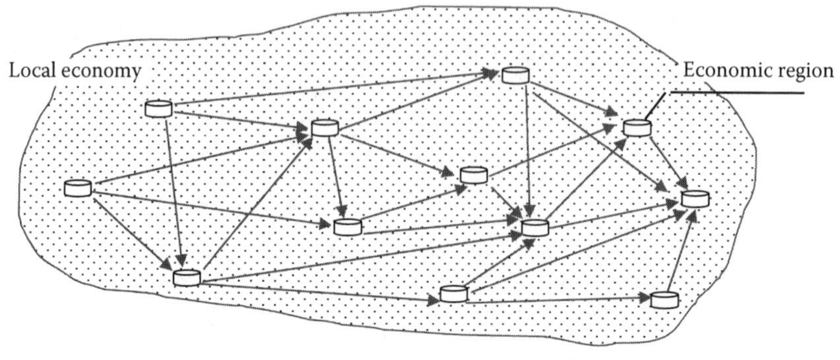

Figure 2.4 The network structure diagram of technology transfer.

flow to this region. In the whole network, any economic region can be one of the main suppliers of technology or may be in need of technology; a region can be an influential member in some areas or a dependent region. The relationships of technology transfer form a dynamic network structure, as shown in Figure 2.4.

2.1.6 Technology Transfer Shows a Positive Link to Technology Level

Technology transfer has always been influenced by the level of technology, from the era of transfer of paper-making and other handicrafts to modern electronic technology transfer as shown in Table 2.1. The table shows that, with the development in the level of technology, the higher the level of technology, the shorter the technology transfer and the faster the speed.

2.1.7 Technology Transfer Is Influenced and Constrained by Many Factors

Jinglian Wu in "creating Chinese own silicon valley" points out that "the high-tech industry development speed of a nation or region is not determined by how much money the government pours into, the number of motivated people, the number of technology developed but whether it has a set of institutional arrangements, the social environment and cultural atmosphere that are conducive to innovation activities and give full play to the human potential."

According to the theory of Maslow's hierarchy of needs, from an analysis of the impact of technology transfer elements, it is clear that technology transfer is mainly influenced and constrained by the introductory systems and incentive policies, security, civilization and honesty, transportation and geographical conditions, the living environment, culture similarity and linguistic affinity, and other

Table 2.1 Some Technology Transfer Recorded in the History of the World

Invention Project	*The Location and Age of the Invention*		*The Location and Age of the Transfer*		*The Transferring Time (Year)*
Paper	China	Second century	Europe	Twelfth century	1000
Powder	China	Ninth century	Europe	Fourteenth century	500
Printing	China	Eleventh century	Europe	Fifteenth century	400
Glasses	Italy	Thirteenth century	Japan	Sixteenth century	300
Mechanical table	Germany	In the early sixteenth century	Japan	In the early seventeenth century	100–150
			China	In the middle sixteenth century	
Lead sulfate room of the legal system	England	1746	Japan	1872	126–186
			China	1932	
The rule of law chlorine bleaching powder	France	1785	Japan	1872	97–134
			China	1909	
Steamship	America	1801	Japan	1855	54–64
			China	1865	

(continued)

Table 2.1 (continued) Some Technology Transfer Recorded in the History of the World

Invention Project	*The Location and Age of the Invention*		*The Location and Age of the Transfer*		*The Transferring Time (Year)*
Cement	England	1821	Japan	1903	82–85
			China	1906	
Railway transportation	England	1825	Japan	1872	47–59
			China	1884	
Match	England	1827	Japan	1876	49–53
			China	1880	
Wire telegraph	America	1844	Japan	1869	25–36
			China	1880	
Open-hearth steel	Germany	1865	Japan	1890	25
Electric light	America	1880	Japan	1890	10
Radio	America	1910	Japan	1925	15–17
			China	1927	
Electron microscope	America	1936	Japan	1942	6
Nylon	America	1938	Japan	1949	11
Semiconductor three-tube	America	1950	Japan	1954	4
Pure oxygen BOF steelmaking	Austria	1953	Japan	1957	4

factors. They are interrelated, interdependent and overlap each other, and arranged in ascending levels of importance from the lowest level to the highest level as shown in Figure 2.5.

Among all the needs, economic interest is the most basic need and the driving force behind technology transfer elements. When preferential policies with regard to technical elements are introduced and a reasonable distribution mechanism is in place in a region, the technical elements will provide a good incentive and reward in this region, and will create an ideal agglomeration effect and will also attract elements of the technology present in the region.

Politeness and honesty are the premise of investment and the guarantee for economic development; without these lawbreakers cannot be controlled. Consequently, these provide protection for productivity transfer.

Transportation and geographical elements are other important factors in the transfer of technology elements. The inconvenience in transport systems and geographical barriers would increase the cost of production, thus affecting the efficiency of investment. With the improved transportation conditions in China,

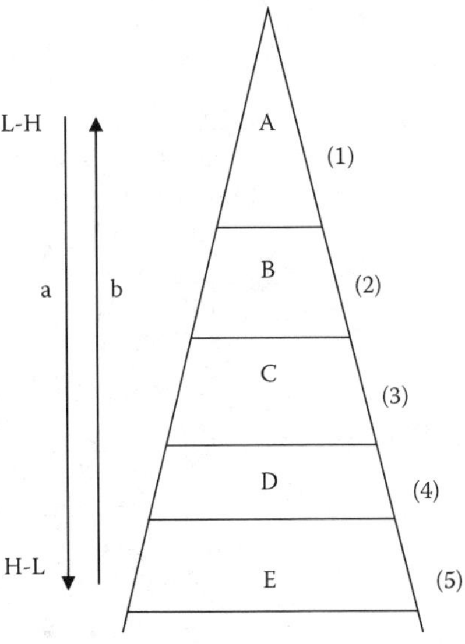

Figure 2.5 Technology transfer factors diagram. A, the needs of self-development B, the needs of living environment; C, the needs of geographic traffic; D, the needs of safe and civilization; E, the needs of economic benefit; 1, the play, development, use of potential; 2, superior environment, comfortable and pleasant; 3, convenient traffic, superior geography; 4, safe and civilization, honesty; 5, with sufficient interest; L-H, low and high; H-L, high and low; a, the density of need; b, the level of need.

the time for transfer has been shortened, reducing the negative impact on the technical elements transfer gradually.

With the development of society and technology, people's demands for a better living environment also increase. Living conditions, facilities, scientific and educational atmosphere, linguistic affinity, and other factors also affect the flow of technical elements.

Once the above requirements are met, the elements of technology transfer will move toward a higher level of development, that is to meet the needs of self-realization—the realization of the technical elements of innovation, to achieve technological progress faster.

2.2 Technology Transfer Properties

Technology's unique features are productivity and transfer. The main features of technology transfer are direction, stage, level of difficulty, "space division," and concentration in the region.

2.2.1 Direction in Transfer

It is generally believed that technology flows from a relatively affluent region to a region of relative scarcity, the direction being directly dependent on price differences. However, the transfer of technical elements can be effected only in relation to other factors. So the technology may not flow to the region with the highest price. The future price expectations of some elements and noneconomic factors (such as political, military, cultural, and natural, etc.) also affect the transfer of technology.

In addition, the features of the technical elements themselves also affect their transfer. For example, technical personnel can be transferred in the international community. However, because of language barriers, cultural differences, and habits as well as government policy restrictions, the transfer is relatively weak although the trend continues to increase. As for the technical elements, the direction of transfer is fixed, namely from technologically advanced countries (or advanced regions) to technologically backward countries (or backward regions). Technology transfer is often closely related to the change in the trade structure of products and their life cycles.

2.2.2 Stages in Transfer

Technology transfer in general has to go through three stages: the influx stage, the establishment of its status in the receiving region, and the receiving region gaining an advantage from the inflow. In the first phase, there is technical outflow from one side to the inflow side. This involves the transfer of the technology elements, and is purely transfer in this stage. In the second phase, the inflow side uses the technology transfer elements in combination with their productivity elements to create a competitive technology. In the third stage, technology transfer elements of the transferees substantially increase, and they even begin to transfer technology out, leading to the development of new technology to achieve a comprehensive upgrading of technology. Only in the third stage, does the technology that has been developed really replace the technology from the transferors. Related technology, management experience, and market channels have also been used to complete the technology upgrade. At this point of time, the first outflow of technology may have compressed or eliminated the technology in this region, replacing it with the new technology with higher technological content.

2.2.3 Difficulties in Transfer

The transfer of technology varies in different industries and regions. The differences are reflected mainly in two aspects—one being the elastic degree of technical control, the other being the sources which are suitable for the productions factors of technology.

In general, the barrier to technology transfer is not high enough. When there is a very slight marginal difference in the technology of two sides, the competition they face is in their relative complete market structure and the demand for the outflow of technology is weak when the technical factors of production such as semifinished materials and knowledge of the environment also match. It will therefore be easier for the transferors to complete the transfer of technology. However, for the technology transfer the two sides are likely to face competition in the market structure and monopoly, and the outflow is to avoid leakage of technology and strictly control the transfer and proliferation of technology.

On the other hand, if the requirements for semifinished materials and knowledge and environmental factors of production are higher, the technical inflow side will have to do a lot of work to develop its own industry and human resources in combination with the technology transfer elements. So, the transfer of technology in this case becomes relatively difficult. The backward region or the technological upgrading of enterprises shows a trend of diminishing marginals. At first, the technology upgrade is easier, but, with the technology gap between the two sides becoming narrower, the difficulties of technical upgrading are expected to increase.

2.2.4 "Space Division" in Transfer

Space division is the term used when one technology is split into many parts and separately transferred to different geographic or business regions. For instance, the technical personnel are always divided by their area of work. The whole production chain, namely R&D, design, training, motherboard production, system production, terminal processing, testing, quality control, etc., is divided on the basis of this technical decomposition; each enterprise only focuses on their own core strengths and advanced resources involved in the technical production chain.

For the main transferors, the meaning of production is limited not only to the manufacturing process, but also to the broad value-added process. In manufacturing, this value-added process includes R&D, manufacturing, sales, after-sales service, and the various segments. To maintain a competitive advantage, transferors not only rely on the complete possession of some techniques, but also look for a comprehensive, comparative, and cooperative advantage, making every effort to participate in and seize the high-tech and high value-added segments, transferring the low-skilled and low value-added link out. At the same time, the development of modern technology makes the development process of high-tech products huge and complex; their requirements for capital, technology, human resources, and management are also getting higher. In the face of such pressure, the transferors must concentrate on developing the core business to reduce their operating costs and risks so that there will be no competition in production.

2.2.5 Regional Agglomeration of Transfer

People in different regions or enterprises gradually understand how to move from comparative advantage to competitive advantage in the process of searching for technology transfer (including the inflow and outflow). Comparative advantage is the static advantage that is determined by party resource endowment and transaction conditions. Competitive advantage is derived from the factors of production, demand, technical supports, and the "aggregated" effect of the resulting formation. Business strategies, government policies, and other "software" factors, provide dynamic advantages. The objective location of each region decides that the transferred technology (including the inflow and outflow) has certain characteristics of the region, and shows a concentration of related industries. The industry concentration, in turn, generates external economies on a scale that will help to improve the competitiveness of the technology to gain a competitive advantage.

2.3 Effects of Technology Transfer

With the globalization of the world economy, technology transfer has become unavoidable. Technology transfer helps to promote optimal distribution of technical resources and provides a strong guarantee for further all-round coordinated development of the national economy and the world's socioeconomic development.

2.3.1 Associated Effect

The theory of dissipative structures and coordination opines that the exchange between system and environment will cause changes in the value of entropy in the system. If the entropy of the output and negative entropy of the input are together more than the sum of the entropy of the input and the entropy generated by the system, the total entropy of the system will reduce and the system will evolve from a low to a high sequence. When the reverse is true, the system will evolve from a high to a low sequence.

A technical system is an open self-organizing system in that both technology transfer and growth constitute the cause and effect and jointly promote the development of the system. In the process of transfer, the flow of technology can bring negative entropy to the receiver (advanced technology enables the receiver to increase production capacity and efficiency), and also provide a positive entropy (for example, technological backwardness leads to reduction in the production capacity of the receiver). When the inflow technology and the technical elements in the region combine, there are two possibilities, one being the positive entropy (if the inflow technology and the regional technology are heterogeneous, they cannot coordinate and combine with technical elements in the region and this causes a decline in the production capacity and efficiency), the other being the negative entropy (the inflow technology and the regional technology are same, they combine with the technical factors to improve capacity and efficiency).

Therefore, in the process of technology transfer, if the sum of the negative entropy inflow and negative entropy generated in the system are more than that of the positive entropy inflow and positive entropy generated in the system, the technical levels of reception will improve, otherwise they will decline.

As shown in Figure 2.6, X stands for the amount of technology transfer, Y stands for the entropy, OA stands for the positive entropy generated by the technical inflow, OA' stands for the negative entropy generated by the technical inflow, OB stands for the positive entropy generated by the combination of technology, OB' stands for the negative entropy generated by the combination of technology. When $|y_1' + y_2'| > y_1 + y_2$, negative entropy is more than positive entropy in the system and the technological levels of reception will improve; when $|y_1' + y_2'| < y_1 + y_2$, the system's positive entropy is greater than the negative entropy and the technological levels of reception will decline.

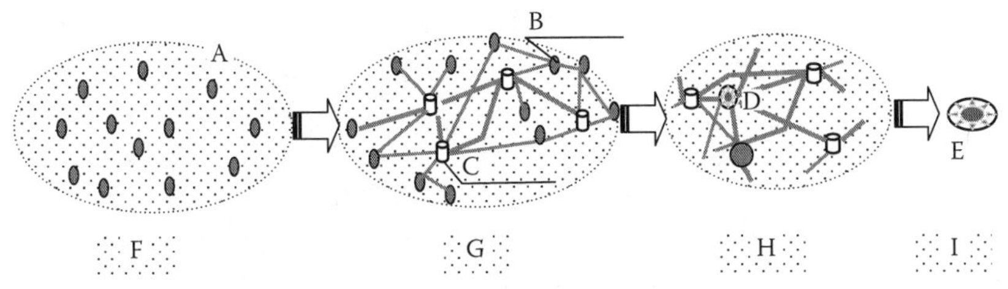

Figure 2.6 Polarization form of technology transfer diagram. A, economic system; B, general domain; C, propelling domain; D, initial polar; E, polar core; F, initial state; G, initial polarized state; H, cluster state; I, polarized state.

2.3.2 Multiplier Effect

There is no doubt that technological progress has dramatically raised productivity and greatly promoted economic growth. In the transfer of technology, there may be different routes of technology transfer such as the "transplant," "variation," and so on in an attempt to achieve amplification in quality and quantity and hence produce the "factor's effect" to achieve an expansion of its own. On one hand, the technical elements flow from the region at a higher technical level to the region at the lower level, whose production efficiency improves and input–output rate increases so that the technology output after the transfer is greater than before; on the other hand, when the technology transfer takes place, it always attaches to some concrete matter (talent, machine, etc.). By the impact of the worker's processing and reprocessing, it will often change the chemical nature of the output and achieve a qualitative change in the output efficiency unlike the productivity growth realized by the direct investment of elements. Economists in the past referred to this as the product of "collaboration" and "comparative advantage" and other horizontal mechanisms of technology. This amplification resulting from the transfer of technology is widespread.

Assuming the benefits of some technical elements in the original resources are y_i and the benefits of inflows are y_i', the multiplier effect of technical elements is shown in Equation 2.1:

$$\theta_i = \frac{y_i'}{y_i} \tag{2.1}$$

2.3.3 Coupling Effect

Coupling is the process that helps to achieve matching, coherent, and coordinated means; achieving coupling is the primary function of technology transfer. Any existing system of regional productivity is bound to be connected with other regions and is not isolated in production activities. Such a connection can be a simple individual communication or a complex combination of multiple communications.

There are two possible scenarios to the coupling effect of technology transfer. First, the purposes of a transfer are not to establish the relationship between the parties (the source of streams and stream places), but to make effective the combination or association of the technical elements of the stream places. The transfer aims to improve the original ecological technology of the stream places. For example, it provides a suitable means of production, effective labor, and so on, in a reasonable combination to improve production capacity and to protect the production activities. However, the production process of the two sides do not form a two-way contact and so the coupling effect in this case is not persistent and is likely to end after only one trading activity. Second, the transfer is critical to the regional input–output chain, the division of labor, and cooperation in the vertical or horizontal direction. Because the transfer forms a two-way contact, the coupling effect is long-term and long-lasting.

2.3.4 Integration Effect

Integration effect is the effect that achieves structural optimization of the internal structure through technology transfer. If the integration is the result of the division of technical activities, then the integration effect is the realization of portfolio optimization (consolidation) on the basis of divided labor. In a micro view, by entering the key elements of the technology (the output from the other side) and by changing the way the elements are associated, the productivity unit will improve to a great extent. From a macro point of view, the region (country) can make use of the horizontal transfer mechanism, against the background of economic division to develop and adjust its leading industries. It can continue to improve the links between industries so as to make the "forward," "backward," and "lateral" transfers help to form an organic whole or network. At the same time, it makes the technology flow in the regional cycle and the international cycle produce the relation of compensation and coordination. The internal and external links are the dynamics of the industrial structure, which contains a structure portfolio with great potential so that the overall economic benefits increase as a result of the "integration" effect.

2.3.5 Polarization Effect

In the early stages of production, because of the abundance or lack of resources, historical and cultural factors, geographical and transportation factors, there was industrial clustering (technical clustering) in certain regions. When the degree of clustering reaches a certain level, cost-sharing of infrastructure, public services, and social management accelerates the improvement in the quality of the labor force and speeds up technological progress and innovation, bringing additional benefits like production advantage. Some of these developed industrial regions rely on their own advantages along with the horizontal adsorption of the surrounding region, leading to a higher concentration of production factors as is shown in Figure 2.7.

The promoting regions (industries or enterprises) that developed (referred to as the East below) have higher productivity; their marginal rate of elements is also high, which has a strong appeal for the region of lower productivity (referred to as the West below). (1) The "polarization" of the transfer fund is the result of limited flow of savings from the West to the East through the intermediary of the capital market; (2) the "polarization" of talent transfer is because the growth of the East is not likely to absorb the unemployed in the West, on the contrary, it has attracted technology and knowledge talent from the West; (3) the manufacturing and export sector in the West will decline because of the competition from the East. In this case, there will be the phenomenon of "the rich regions get richer, the poor regions become poorer," and produce the so-called polarization effect.

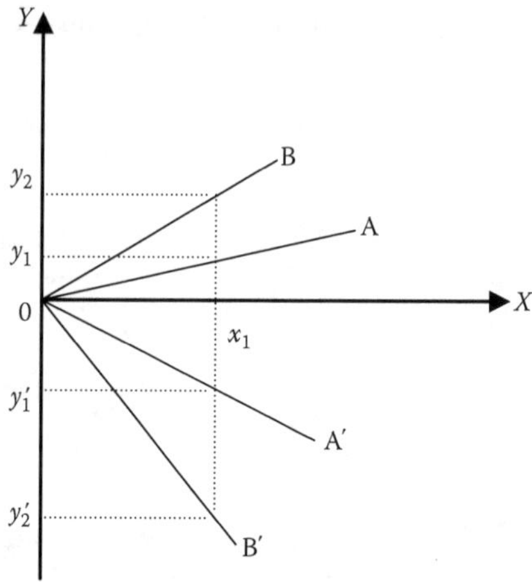

Figure 2.7 The associated effects of technology transfer.

2.3.6 Equilibrium Effect

The transfer of interregional technology can promote common economic growth and the complementary optimization of industrial structure so as to narrow the gap between the regions. Economic prosperity in developed areas will continue to expand the demand of primary products in the West. At this time, if the supply elasticity of the primary products in the West is large enough and can meet the need of the East, the economics of the Eastern and Western districts will grow further; their growth rate depends on the conditions of trade, that is, the price index ratio of interregional exports to the imported goods. If supply elasticity of the West is lower, the demand for the product is more than the supply, leading to a rise in the price, which may promote the growth of production in the West and the improvement in the conditions of pay. On the other hand, the growth of goods in the East will slow down because the price of goods increases so that the gap between East and West narrows gradually. At the same time, the technology transfer between East and West can not only bring about the West-led growth but also promote the structural changes in the West. The East will follow the development of the West, integrate with it, and begin to accept the division and transfer of industries in the advanced regions, merging into the circle of economic development in the East.

2.3.7 Spillover Effect

Sharp, Register, and Gelimisi, in their book *Economics of Social Problems*, defined the social spillover effect of consumption when the consumption or services of products leads the change in the degree of satisfaction for the consumers. Technology transfer will not only generate effects in the transfer but also generate effects in the other social facets such as the cooperation between companies, the spillover effects of personnel transfer.

Assuming one company has the advantage of a certain technology or information, when it contacts the suppliers or customers, they may be on a "free ride" from the company's advanced products, processes, or technical market knowledge so that the spillover effect takes place. Even if the company charges a fixed fee from local suppliers or customers, it cannot gain from the technological progress brought in by local manufacturers.

The transfer of human capital is one example of the spillover effects of technology, absorbed by local enterprise as a result of employees' transfer. When the human capital of region A (industry or enterprise) is transferred to region B (industry or enterprise), the technology that is mastered by the human capital in region A is naturally used in region B.

As market economy is competitive, technology transfer is not only the inevitable result of the competition, but also a reflection of the market efficiency. Technology transfer has a profound impact on the economy of a country or region whether there is inflow or outflow. It is important to understand the positive effects of technology transfer and make good use of it. At the same time, it is necessary to take the negative effects of technology transfer into consideration in the research on the questions in technology transfer. Only then can we make good use of the positive effects of technology transfer, avoid its negative effects, and achieve a benign technology transfer.

2.4 Model of Technology Transfer

Technology is a social artificial system; the transfer of the elements is not entirely spontaneous. It depends on certain social conditions, with several parties (the state, enterprises, and other economic entities) involved in the scope of activities. In other words, the technology is transferred only in certain modes.

2.4.1 Divided by the Actionable Mechanism

In the real economic environment, transfer of technology is actually driven and controlled by two forces, namely market forces and plan forces. Therefore technology transfer is divided into two models: the market-guided control transfer and plan-guided control transfer.

2.4.1.1 Market-Guided Control Transfer

Market force is dependent on the law of value and the mechanism of supply and demand, which is influenced by changes in the prices, interest rates, wages, and other valuable "leverage" factors to influence, guide, and regulate the technology transfer. The sequence of interests, which are the dynamic factors of technology transfer, is quantitatively reflected by the size of the returning rate of transfer. Market force impacts the role of technology transfer through the interregional trade, interdistrict collaboration, and inter-district-specific investments.

The market-guided control technology transfer is purely based on economic reasons. In a sense, the technology transfer is a response to the local search for better opportunities, benefits, and economic space in the market. The key factors that restrict such transfer include the existence of a more complete and developed unified market system, including the backward regions, which is a necessary condition for reasonable and adequate transfer of technical elements, the existence of an effective mechanism for accepting and responding to the transfer in productive systems for the market signals; the comparison between the expected interests and existing interests of the element's transfer for the parties; the transfer of the innovative spirit of the parties, as well as the actual ability to overcome the market risk of transfer, the psychological adaptability, and so on.

2.4.1.2 Plan-Guided Control Transfer

The plan-guided control technology transfer is related to market-guided control. The transfer is under the guidance of the domestic economy, social development strategy, and industrial policy, with the operational plans and programs giving guidance and regulation to technology transfer. The plan-guided control technology transfer is a planned transfer of orientation; "proliferation," "focus," and "injection" are the three common forms of operation. For example, China is presently focusing more on the industrial zone, favoring investment in some regions, interregional migration of industries, and organizing collaboration and production, reflecting the role and requirements of the plan.

Compared to market-guided control technology transfer, plan-guided control technology transfer has, in addition to economic reasons and motives, noneconomic reasons and motives, such as the political environment, economic unity and national security, social equality, national unity, and reducing regional differences. For a particular developing socialist country, the planning control technology transfer is an important and necessary component to combine a planned economy with market regulation.

2.4.2 Divided by the Span of Transfer

Technology transfers occur between and among developed and developing countries (regions). Even as there are differences in technical levels between the countries (regions) of transfer, the spans of technology transfer are also different, so are the vertical transfers and horizontal transfers of the technology.

2.4.2.1 Vertical Transfer

In accordance with the international division theory and gradient theory, the vertical transfer (also known as vertical movement) refers to the transfer from the developed countries (developed regions of technology) as the supply side of technology (the source of technical streams) to the developing countries (less developed regions of technology) as the accepting side of technology (the technical stream places). In general, the developed level of technology or the technology gradient is higher in the advanced countries whereas the developed level of technology or the technology gradient is lower in the developing countries. The technology transfer that is from the higher level of technical development (higher gradient) to the lower level (lower gradient) is known as vertical transfer or vertical movement; in reality, it is the transfer of technology from developed to developing countries.

2.4.2.2 Horizontal Transfer

The horizontal transfer (also known as horizontal movement), refers mainly to the technology transfer between the countries and regions which are at the same or similar levels of technological development (gradient).

Vertical movement in the social scene can make people hope for improvement in the social situation. Horizontal movement in the social scene raises the hope of vertical transfer after a change in the social environment. For any enterprise or institution, no location can be fixed. They should be allowed to compete fully, for the potential of technology can be maximized only in the face of sustained, full competition.

Some foreign scholars, such as E. Mansfield of the United States, have defined vertical and horizontal movements as follows. Vertical movement is the transfer of the basic scientific research of country A to the application science of country B, or the transfer of achievements of the application science in country A to production in country B. Horizontal transfer refers to the transfer of new technologies applied in country A to the fields of production in country B. These definitions and divisions have theoretical and reference values also.

2.4.3 Division by the Vector of Transfer

2.4.3.1 Transfer of Physical Elements

The transfer of physical elements refers to the transfer of complete sets of equipment, tools, constructions, and other physical forms. These physical forms of Technology, "Intermediate Technology", constitute most of the transfers. This technology is likely to be obsolete soon as it is embodied in a physical form. Once developed countries have a new technology they will try to transfer the relatively backward technical equipment abroad to shift the "intangible losses" of technology. At the same time, the demand "Intermediate Technology" is great in most developing countries because of poor absorption capacity in the domestic market. This is the basic background against which technology transfers from developed countries to developing ones happen.

2.4.3.2 Transfer of Information

The transfer of information refers to the technology transfer through patents, technical know-how, design, formulation, and other forms of knowledge. Such technology is not immediately used in production in general when it is usually in the technical testing stage. In general, the mutual transfer of this type technology in developed counties is larger, and forms the technology trade wherever permitted.

2.4.3.3 Transfer of Capability

The transfer of capability is also known as the transfer of "technology-mind," and is the transfer of technical talent because the technical talent is the "living" carrier of ability in technology creation so that the transfer is equivalent to the transfer of technology to create (invent and develop). Technical talent is the basis of technical development and transfer of international technical talent is the most common having a positive effect in promoting the development of all technology. Developing countries need more technical personnel exchanges, however the exchange of technical personnel in general flows from developing countries to the developed countries rather than in the reverse order; the talent transfer because of immigration is one such. Such an irrational structure in the transfer of talent is in the forefront of the grave problems of developing countries. It will hopefully be reversed by the cooperation between developing and developed countries.

2.4.4 Division by Absorption and Processing

This grouping is based mainly on Mira Wilkins' classification, and it deals with the follow-up of transfer of technology.

2.4.4.1 Simple Transfer

Simple transfer refers to a onetime technology transfer, which is used directly by the recipient and the supplier does not question the recipient copying them. This is equivalent to the transplant of technology or of industry, and it is simply a "used" way.

2.4.4.2 Absorption Transfer

Absorption transfer refers to the technology transfer with a follow-up process of absorption. The follow-up process of absorption includes imitation, duplication, and improvement of the technology.

A simple transfer of technology is the basis of absorption transfer, but from the recipients' point of view, subject to certain shortcomings or limitations, to the effectiveness of the technology. Once the imported equipment is damaged, it has to be imported again because it cannot be duplicated. In contrast, the transfer of technology, which can be absorbed, has the advantage over the former. As a result of its follow-up, domestic absorption, the effectiveness of the imported technology is extended. For example, after the 1960s, simple transfer is not the mainstream in Japan; when the first machine is imported, the second one must be produced domestically for more benefits. In fact, in developing countries the more successful imported technology is generally a combination of a simple transfer and an absorption transfer; the proportion of absorption transfer is gradually increased. If there is a gap in a certain field of technology or program, it could simply be imported to avoid the risk of developing one and to save the cost of development; for those domestic-based, but still relatively backward technologies, there should be an absorption transfer, that is, digestion and absorption to achieve the following aims: on the one hand, the technology need not be imported again and again; on the other hand, imported technology and domestic technology could be combined systematically.

2.4.5 Spatial Model of Technology Transfer

2.4.5.1 Types of Spatial Transfer and Its Disciplines

The basic models of technology transfer are the epidemic model, the skip model, and the hierarchical model. Companies are the subjects of flow technology, which is largely dependent on enterprise technology strategy, including the proliferation of conditions, means, and opportunity.

In view of the resource conditions, the motivation to lower costs encourages technology transfer from advanced countries to less advanced ones; this is the skip transfer model with epidemic transfer to remote areas. Technology life span comprises four stages–technology application period, growing period, maturing period, and standardizing period. The application and growing period are concentrated mainly in metropolitan cities. In the mature period, companies transfer to moderately developed countries whose markets have great potential whereas in the standardized period, technology transfers into developing countries where the cost is low, shaping the model from developed to developing countries via moderately developed countries. When such transfer occurs, the transfer of technology becomes more widespread owing to the great demand and technology resources mature. Such examples can be found in the introduction of 138 manufacturing lines in television sets in China in the early 1980s, which led to the establishment of television companies in every province. From the market point of view, enterprises, when entering new markets, tend to transfer technology to remote countries and regions to avoid competition with leading companies shaping the skip model of technology transfer.

Among the many means of technology scattering, technology transfer is often chosen to take place in some remote countries and regions to keep technology confidential, forming the skip model. Advanced large scale corporations often develop their accessory branches in neighboring areas to keep the connections, which show the epidemic model. The modeling function is the main pattern to penetrate local technology for Foreign Direct Investment (FDI), which decays with distance and shows the need for the epidemic model. The value of cost and market in international ventures leads to the skip model of technical modeling. Such models can also be seen in the application of new measures resulting from an up-front connection with dealers in capturing new markets through international ventures.

2.4.5.2 Case Study: Spatial Pattern of Technology Transfer in Zhujiang Delta Region

Considering companies such as Jinpeng, Besta, Motorola, Hualing, and TCL as examples, these modes of cooperation can be regarded as skip models not only geographically but also on the level of techniques. Moreover, two kinds of new space-time models are represented in companies such as Hualing and TCL.

Hualing Corporation chooses a time-space model that is moved from a remote area to a surrounding one, introducing new techniques from developed countries and combining them with local techniques. Such a model is a combination of the skip and epidemic models in technology transfer. At the outset, in ignorance of techniques in air conditioners, Hualing cooperated with the Japan-based Mitsubishi Corporation that acquired the most advanced techniques of air conditioners and afterward developed production ability quickly, a model of skip development. Thereafter, to reduce cost, Hualing placed greater emphasis on self-innovation and cooperating with local universities aims at making use of the research capacity in colleges, to obtain techniques and reduce cost initially as well as develop scattering techniques in remote regions as the transition from epidemic model to skip model. At the beginning of development, TCL chose the neighboring companies to obtain techniques and reduce cost by means of outsourcing and merging resources. As it grew, with distance no longer a constraint, the corporation expanded to north China and Indo-Asia Penn shaping the surrounding-remote model.

Chapter 3

Dynamic Mechanism and Measurement of Technology Transfer

Technology transfer is based on the nature of the transfer participants, is supported by the social, economic, and cultural environment, and driven by the technical potential difference; the transfer of information and material is the process. It is a complex process comprising many factors, such as technology, economy, society, enterprises, information, culture, etc. All the elements that affect technology transfer are interrelated and interact, and their characteristics as well as the interactions determine the time-space development model of technology transfer. Meanwhile, the process of transfer will in turn influence the changes in and relationships between the factors. It is clear that the technology transfer process is a dynamic network of interrelations and interactions between every affecting factor and the transferring subject, between the process and the transferring environment.

3.1 Constituents of the Technology Transfer System

3.1.1 Hierarchical Structure of the Technology Transfer System

Bertalanfy, the founder of system science, believed that the theory of hierarchy is an important pillar of the general system theory. Every system possesses many different levels in which the higher levels manage the lower ones, i.e., the latter are subordinate to the former; in other words the former dominates the latter and the lower levels obey and support the higher levels. Hierarchy plays the same role in the system of technology transfer, which can be divided into three levels namely interregional flow system, regional flow system, and international flow system as shown in Figure 3.1.

The technology transfer systems with different levels have different evolution laws. For example, regional transfer of technology is certainly different from the transfer between two enterprises in the same city. The former transfer depends mainly on the interaction space in the transferring regions, whereas the latter is determined mainly by the technology situation, technology management, and communication of information between two enterprises.

Technology transfer can also be divided into macro- and microlevel transfer. The macrolevel transfer mainly studies the evolution of technology transfer in the transfer space to reveal the cause and mechanism of the unbalanced development of innovation in technology transfer in space, to improve and optimize the transfer process, to find ways to expand the application of the innovative technology to obtain the potential economic benefits of the technology. On the other hand, the microlevel transfer mainly focuses on the technical proliferation between main adopters and analyzes the realization process and the law of transfer in potential users to reveal the micromechanism of technology transfer. Thus the strategies for improving the micromechanism of innovative technology transfer among potential adopters can be found, which will enhance the efficiency of the transfer process and promote economic development.

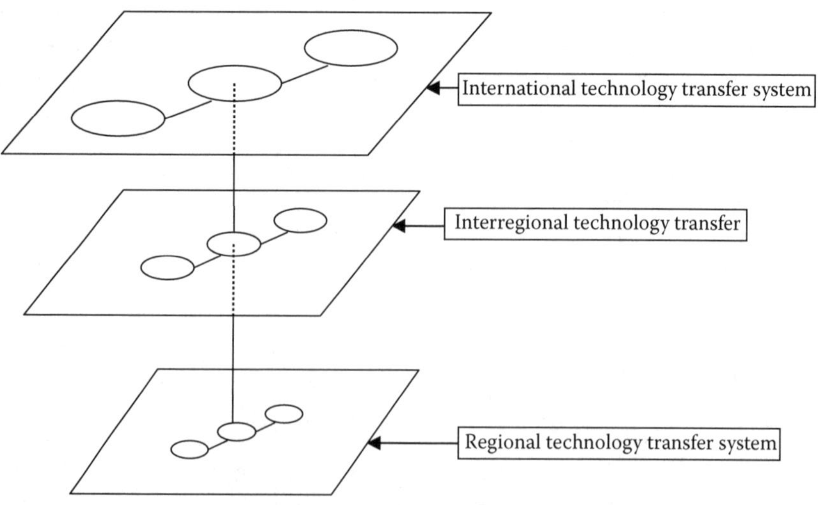

Figure 3.1 Hierarchical structure of technology transfer system.

3.1.2 Basic Composition of the Technology Transfer System

Technology transfer is a complex technological, economic and social system. The problems of why, where, how, and who to transfer the technology are all related to this system. The system consists of the object of technology transfer, the source and the sink of transfer, the information channel, the auxiliary system, and the incentive mechanism, and so on, as shown in Figure 3.2.

3.1.2.1 Source of Technology Transfer

The source of technology transfer refers to the enterprises or organizations which spread the technology, it is the space performance of the technology output side. Sometimes the location of technology is referred to as the source of technology transfer. The technology export of the transfer source integrates with the technology of the owners who generally approach organizations possessing and innovating technologies like enterprises, research institutes, the intermediaries, relevant government departments, and so on. These owners influence the process, the speed as well as the quality and the scope of technology transfer through the transferring behavior.

3.1.2.2 Object of Transfer

This refers to the technological innovation being transmitted and promoted in the transfer process. In a transfer system, the connotation of transfer object may be

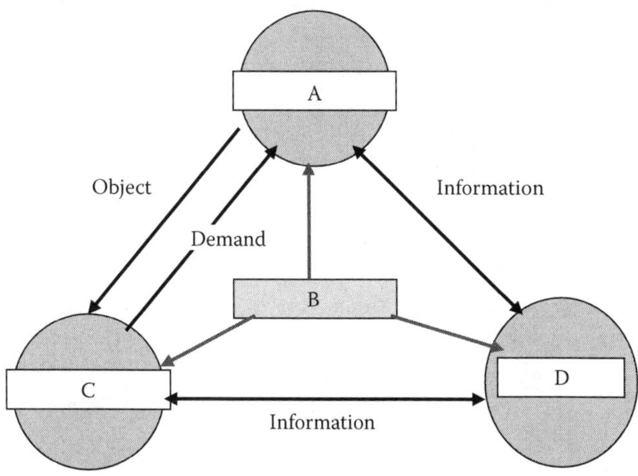

Figure 3.2 The technology transfer system diagram. (A) The source of technology transfer. (B) The incentive mechanism. (C) The sink of technology transfer. (D) The auxiliary system.

production innovation, process innovation, marketing innovation, raw materials innovation, organization, knowledge innovation, etc. The innovation technology transferred is often in the form of tangible technology such as equipment, product, and other intangible technologies like software, documents, and disks.

3.1.2.3 Sink of Transfer

The sink of transfer refers to the potential of enterprises to absorb and adopt innovative technology; sometimes the inflow of innovative technology is also known as the sink of transfer. The integration of the innovative technology by the adopter is the introductory action in the technology transfer system. The adopters' evaluation of and decision-making on innovative technology will affect the process, speed, quality, and scope of the transfer. Therefore their adoption behavior is taken as a necessary condition and microbasis for the technology transfer whereas their quality features are the dominant factors that affect the realization of innovative technology transfer.

3.1.2.4 Information Channel

Information channel is the basis of technology transfer, acting as a bridge for the output and demand of technology. Earlier information exchange activities were organized by the government. The so-called action of setting up stages for companies and research institutions actually means that the government sets up the information channel to facilitate the communication and cooperation between the two sides. To strengthen bilateral exchanges, the government also arranges for researchers from research institutes to visit and investigate the companies looking for cooperation opportunities. At the same time organization managers and technology directors will pay a visit to research institutes to learn about scientific achievements. There are several types of information channels, such as exhibitions, achievement shows, the fair exchanges, technical seminars, study tours, popular science lectures, and so on. Entering the era of computer networks, information networks is a modern information channel and many scientific research institutes have established their own Web sites to publicize scientific achievements and technical information. Many companies introduce the manufacturing and production information on their web sites and local governments and a number of technical intermediary units also use the information network to establish the online technology market. The information channel of technology transfer has enhanced mutual understanding and exchange between the owners of technology and the demand-side, and promoted technology transfer.

3.1.2.5 Auxiliary System

As a support, the auxiliary system provides a material guarantee for technology transfer to facilitate the commercialization of technology transfer. It is a complex and long process to transfer technology to production for practical use to meet the market requirement.

In this long process there are many technical, management, and economic issues to be solved, requiring the improvement of the technical process and its stability, the expansion of the production scale, high quality, and output. The issues also include developing new markets, exploring the feasibility of the economy, and so on. Is therefore clear that it is not only a big investment with a big risk but also relates to many factors such as human, material, finance, resources, etc. The basic function of the auxiliary system is to support the development and the commercialization of technology making it finally applicable to products and services. The technology transfer auxiliary system has several forms; it can be pilot-plant bases, incubators and technology parks, institutions to promote productivity, technological innovation institutions, technology transfer agencies, engineering research institutions and technology intermediaries, and in some cases it is the integrated form of some or all of the above. Through technical exchanges, technology development and product design, conciseness of problems, technical advice, and others, this system creates a good environment and opportunity for technology transfer and supports the commercialization of technology. In this information age, the virtual R&D bases, the service centers of technology information and transfer, have become a modern and effective kind of auxiliary system.

3.1.2.6 Incentive Mechanism

Incentive mechanism is a key factor in improving the effectiveness of technology transfer. As a catalyst, it can enhance the functions of many factors in the process of technology transfer and speed up the process. During technology transfer, technology owners, the technical staff, and technical needs are the most relevant factors, and the incentive mechanism is just intended to speed up technology transfer and improve the efficiency by improving the atmosphere, helping create innovative management systems, coordinating relations between subjects and objects, stimulating the technology transfer initiative, and encouraging the transformation of scientific and technological achievements. Meanwhile by improving the structures of enterprises and institutions, the owners will develop technology based on the demand. In this way the speed and efficiency of technology transfer will be enhanced.

3.1.2.7 Internal Environment of Technology Transfer

3.1.2.7.1 Meaning of the Internal Environment of Technology Transfer

The internal environment of technology transfer refers to a collection of all the ubiquitous factors that are the basic components of the system, and is different from the external environment beyond the boundaries of the transfer system. The external environment, which consists of the economic mechanism, political, legal, and social environment, has an indirect impact on the diffusion process and can be treated as the external and fixed factor. The internal environment of technology transfer includes the technical base and the structure of the transfer space, market structure, industrial structure, resource endowment, infrastructure, and other economic conditions. These are the basic conditions of the transfer process, which is in a dynamic state of evolution.

3.1.2.7.2 Function of the Internal Environment of Technology Transfer

It is a typical interacting mechanism between the transfer process of innovation technology and the internal environment of technology transfer. On one hand, the transfer process is impacted and constrained by the internal environment of transfer. On the other hand, it acts on the internal environment of transfer to further its development and evolution; then the changed environment will affect the transfer process so that a new pattern of movement is formed. The new pattern in turn influences the internal environment of technology transfer, and the cycle goes on. This way that the technology transfer process moves forward in the form of interaction with the internal environment of transfer.

The internal environment of technology transfer has a dual effect on the transfer process. The former supports the development process but has some limitations and obstructive factors. Therefore the adopting process of potential users becomes time-consuming and makes the transfer a limited growth process.

3.2 Dynamic Mechanism of Technology Transfer System

3.2.1 Analysis of the Technology Transfer Affecting Factors

3.2.1.1 Technology Potential

Technical potential refers to the technology level in a specific area. Shen Yue, Fu Jiaji, and other scholars believe that the potential between creators of technology and the surrounding area is caused by technology innovation. To eliminate this difference, creators will be forced to diffuse the technology around or make the surrounding regions learn and imitate the innovation. Through technology transfer, information exchange, personnel transfer, international trade and investment, and other technical means, technology transfer may take place between groups of people, among enterprises or regions. The greater the potential difference is, the higher the transfer conditions and the more difficult it is for technology transfer to take place. On the contrary, if the conditions are relatively low, technology transfer is more likely to take place; thus the characteristics of technology transfer duality are apparent. Moreover, technology potential differs at different scales of space, for example skip transfer exists in a macroperspective and expansion transfer and gradient transfer are common in a micro perspective.

3.2.1.2 Space Distance

Space is the most important factor influencing technology transfer in a microscale. The interaction, limitation, and mutual promotion of potential technology adopters will affect the transfer of elements (information, conceptions, materials, talents, capital) in space and technology transfer directly. To be specific, the transfer process starts from the communication of information by the innovator to the initial receiver. The strength of technology potential decreases with the increase in distance. This makes distance one of the most important factors that affect the transfer process. The closer the receiver enterprise is to the technology source, the higher the possibility that it can receive the technology and vice versa. Therefore, on a microscale, distance requires space agglomeration to reduce the time and cost of introducing technology and technology cooperation. Silicon Valley, Tsukuba, Japan, Bangalore (India), as well as Hsinchu, Taiwan, Beijing's Zhongguancun Hi-Tech Park, and other well-known domestic and international developments show that high-tech enterprises agglomerate mainly in areas such as science and education, special economic zones and economic and technological development zones, industrial zones, and the new areas.

3.2.1.3 Transfer Channel

This concept refers to the internal and external environment of technology transfer. The external environment of technology transfer includes regional economy, policy, laws, and social environment whereas the internal environment refers to the regional system consisting of economic development, technological structure, market structure, industrial structure, the flow system of information, resource endowments, and enterprise quality. The smooth technology transfer channel should be conducive to technology transfer, otherwise it will impede the transfer. The technology transfer channel features are dynamic. With differences in the economic system, industry and market structure, social and culture environment, the countries or regions with the same technology potential may shows different preferences to technology exploration and introduction resulting in different strategies to develop technology and thus the transfer refers to different contents and levels.

Therefore the transfer of technology is under the comprehensive influence of the technology potential, distance, and the technology channel. Specifically, the technology potential difference is the fundamental driving force and the distance is an important factor affecting cost whereas the existence of the technology channel is a necessary condition to achieve technology transfer. Domestic and foreign regional development show that a high-technology potential, and short distances and convenient access between technology partners, help the smooth transfer of technology.

3.2.2 Constituents of the Dynamic Mechanism of Technology Transfer

The view of Japanese economist Teng Chaiyou on the driving force of technology transfer is very clear. He believes that interaction between the needs and reserves and resources (including human resources, capital, equipment, information, etc.) is necessary to meet the demands and promotes the development and application of technology. However, complementary interaction between one country's N–R (Need–Resource) relation with that of the other country can achieve effective transfer of technology. Complementary interaction also determines a country's industrial development and its foreign economic activities. If R is insufficient in a country, the bottlenecks that appear will promote technological innovation. If it is better for a country to import technology rather than to carry out technological innovation on its own their N–R relationship becomes the N–R relationship of importing technology, whereas in the other country the relationship becomes the N–R relationship of technology transfer. Such complement interactions of N–R relations between two countries includes the demand and resource transfer, information exchange, diversified technology transfer channel, system and infrastructure of technology transfer (including the patent system, technical education, training system, and other institutional factors, as well as transportation, communication facilities, research, and development institutions, universities, and other hardware factors), and so on. The more extensive and active the N–R international relations, the easier it will be for the transfer of technology.

The process of technology transfer can be taken as the gradual entry of technological enterprises with potential demand, whereas its dynamic mechanism can be divided into the following three aspects: the pulling mechanism, thrust mechanism, and coupling mechanism of technology transfer.

With regard to the pulling mechanism of technology transfer (that is the dynamic mechanism of the demand party), the innovation–imitation theory represented by Schumpeter (1912), Mehret Ridge (J1Merenish), and Mansfield (E. Mansfield) argues that the reason for the entry of new enterprises is subject to the induction of a profit monopoly. When an innovative technology is not yet fully diffused, the enterprise that has adopted the technology will produce a short-term balance in monopolistic competition; because the price is higher than the average cost at this time, the enterprise will win excess monopoly profit. In practice, this could continue to attract new enterprises to enter. In the early period of innovation and technology transfer, due to excess monopoly profit, innovative enterprises are often reluctant to carry out technology transfer; however, as a result of the transfer fee and the risk of complete monopoly, innovative technology suppliers are forced to diffuse so as to promote technology transfer. Finally the price will equal the average cost, when excess monopoly profit will become zero and no more new firms will be willing to enter and a long-term balance "that is a state of full diffusion" will be reached and the transfer activity will come to an end.

The theory of production cycle discusses the thrust mechanism of technology transfer (that is the mechanism of innovative supply) from the point of view of international technology transfer. It points out that fast transfer is not good for enterprises with innovative technology when considering the maximum profit. The enterprises will determine when to diffuse the technology at different stages of its life cycle. In general, these enterprises will first export products of the new technology. In the process of export, with the products growing in the local market, profits also decrease. Meanwhile the product is gradually adapted to local conditions so that the local factors can also produce the product. Thus, enterprises in turn make direct investments to achieve profit. Finally they begin to export technology directly, moving the integral technology and product process to that country or region, which is the realization of the diffusion of innovative technology.

Generally it is believed that the reason for technology transfer is the technological difference between the demand and supply sides, and the external condition is the coupling relationship of demand and resources of the two sides. And the internal dynamic mechanism is the supplier choosing or delaying the diffusion at different stages for maximum profit, with the demand-side introducing innovative technology for additional profit.

3.2.3 Force of Technology Transfer System

According to the theory of technology transfer, because a new technology can improve efficiency and create more value or can save labor costs and improve the system thus creating new markets, there is "potential difference" between the innovator and the surrounding space. As a result of the existence of the "potential difference," a "field," called the field of technology transfer, develops. According to the theory of field, a power balance will promote the diffusion of innovative technology or encourage surrounding areas to study and imitate the technology to eliminate differences, in which case the transfer will take place objectively. In the field of technology transfer, when the sink exists, the field will have an effect on it to promote the introduction of new technologies. Theoretically, the action of transfer has the following two reasons, first, the transfer is driven by internal and external pressure and second, the sink of transfer is attracted to accept the technology transferred.

3.2.3.1 Attraction of Transfer Sink for Technology

The direction of the transfer option depends on the gravitation field of the transfer sink. The following factors influence the direction of transfer:

1. Labor force, including the price, technical level, technical ability, etc. of labor.
2. Resources—regions with abundant resources have greater attraction to technologies.

3. Location—regions geographically close to the source of transfer is more likely to receive the technology.
4. Marketing factors—a developed market is an important factor to generate high profits. Therefore expanding markets of commodity and technology provide the right conditions for the development and further innovation of the technology.
5. Financial situation—the diffusion process is often accompanied by investment behavior; especially the diffusion of high-tech processes. Well-developed capital markets have an important impact on the source's adoption of the technical achievement.
6. Institutional factors, including the policy of the source, the legal environment, protection of intellectual property rights, and environmental risks.
7. Attraction of technology—this attraction has a great impact on technology transfer, especially when technologies have increasingly become the dominant factor of productivity. Technology attracting technology is a more advanced method of technology transfer.

To sum up, these factors, namely labor, resources, location, market, capital, technology, etc. have a direct influence on the transfer. Therefore, the following model can be used to represent the attraction of the transfer sink to the technical innovation, that is

$$V = \left(X_1, X_2, X_3, X_4, X_5, X_6, X_7\right) \tag{3.1}$$

where

V represents the attraction of the transfer sink to the technology
X_1 is the state of the labor force
X_2 is the state resources
X_3 is the location factors
X_4 is the market factors
X_5 is the financial situation of the factors
X_6 is the institutional factors
X_7 is the attractive technical factors

As in the potential field in mechanics, the gravity gradient function of the technical transfer field can be figured out:

$$X = -\left(\frac{\partial V}{\partial X_1}, \frac{\partial V}{\partial X_2}, \frac{\partial V}{\partial X_3}, \frac{\partial V}{\partial X_4}, \frac{\partial V}{\partial X_5}, \frac{\partial V}{\partial X_6}, \frac{\partial V}{\partial X_7}\right) \tag{3.2}$$

The different distributions of the field of every factor influence the field in different ways; regions will differ in states and form different levels of gradient. The diffusion of technology generally follows the diffusion law of gradient and spreads toward the direction with smallest gradient and with the largest gravitation.

3.2.3.2 Impetus for Technology Transfer of the Sink

The impetus of the technology transfer is affected by the following factors:

1. The cooperation intention based on the development strategy of the transfer source.
2. Market factors, that is, the prospect of the transfer product in the market of transfer sink.
3. The technical ability of the transfer source.
4. System factors, including the estimation and judgment of the source on the transfer sink in terms of the society, politics, economic environment, and containing the policy and legal environment of the transfer source, national security policy, environment risks, etc.
5. The source's choice of transfer time and type of technology.
6. Excellent geographical location near the transfer sink can send out more radiation to the sink.
7. Options in the transfer mode of the transfer source (joint-venture, cooperation, sole proprietorship, etc.).

On the lines of the definition of the sink's attraction to the technology, the impetus of technology transfer from the source is defined as follows:

$$P = P(y_1, y_2, y_3, y_4, y_5, y_6, y_7) \tag{3.3}$$

where

P is the impetus of technology transfer of the transfer source
y_1 is the cooperation intention based on the development strategy of the transfer source
y_2 is the market factors
y_3 is the technical ability of the transfer source
y_4 is the system factors
y_5 is the transfer time and type of technology
y_6 is the location factor
y_7 is the transfer mode of the transfer source (joint-venture, cooperation, sole proprietorship, etc.)

3.2.3.3 Resistance of Technical Transfer Field

Technology transfer is impeded by the following environment factors:

1. Barriers of information delivery
2. Uncertainty of the benefits of innovative technology
3. Incompatibility

4. Adopted price
5. The space constraints and so on

We define the block degree of the technical transfer field transfer as follows:

$$R = R(Z_1, Z_2, Z_3, Z_4, Z_5) \tag{3.4}$$

where
R is the block degree of technical transfer field
Z_1 is the block of information delivery
Z_2 is uncertainty of the benefits of innovative technology
Z_3 is the incompatibility
Z_4 is the adopted price
Z_5 is the constraint of space, etc.

3.2.3.4 Gravitation of Technical Transfer Field

The integration of the gravitation of the transfer sink and the impetus of the transfer source acting on the innovative technology is called the gravitation of the technical transfer field, as shown in Formula 3.5.

$$F = V + P \tag{3.5}$$

F represents the gravitation of the technology transfer field.

According to the law of universal gravitation, the gravity is directly proportional to the quality of two objects and inversely proportional to the square of the distance between the two objects. It is the same case in the technology transfer field.

We divide F into the two parts; one is a comparative superior gravity whereas the other is the demand gravity, as illustrated in the following formula.

$$F = \alpha F_1 + \beta F_2 \tag{3.6}$$

where
F_1 is known as the comparative superior gravity
F_2 is called the demand gravity

$$F_1 = \frac{C_1 M_1 M_2}{R^2} \tag{3.7}$$

$$F_2 = \frac{(M_1 Q_1 - M_2 Q_2) M_1 M_2}{R^2} \tag{3.8}$$

where

C_1 represents the comparative superiority, which is given by the ratio between r_t and r

r_t is the average profit rate of the technical adopter at time t

r is the regional average profit rate of the potential adopters in regions that adopt the technologies

M_1 is the supply quantity of the transfer source

M_2 is the number of potential technology adopters of the transfer source

Q_1 is average economic scale of technology adopters

Q_2 is average economic scale of technology adopters of the transfer source

R is the block degree of technical transfer field

α, β are the coefficient of the gravitation components

Because of the field gradient, changes in the field elements arise during the technology cross-transfer. The source with outward technology diffusion is called a positive source, technology flowing into the transfer sink is a negative source. Given a block degree for transfer, more technology providers will bring about larger gradient than the surrounding area and greater strength of transfer for the source, whereas for the transfer sink, more potential users will generate more demand as well as greater absorption strength.

The comparative superior gravity and demand gravity result in technology transfer. As the block degree R is negatively quadratically related with comparative superior gravity F_1 and demand gravity F_2, small changes in R will cause larger variations in F_1 and F_2. When there is a great difference between the source and the sink of technology transfer in the characteristics of field, stock degree will be more, which goes against technology transfer. On the contrary, technology transfer will be easily carried out when they are similar.

3.3 Imitation of Technical Innovation and Transfer Model in Two Regions

The relationship between technology transfer and technological innovation is essential for the analysis of economic growth. If the technology leader or technology innovator cannot get profit from the progress achieved in the application of the technologies, then the process of technical innovation and transfer will yield a decreasing profit and the innovation speed slows down thus affecting the technology transfer process. Effective transfer, when improving the technical level in the region where the technology is introduced, will reduce or raise the expectation of the technology providers and will affect the next generation of technology innovation. Therefore, the research into the relationship between technology innovation and technology transfer becomes the premise for analyzing technology transfer and economic growth.

3.3.1 Assumptions

Assumption 1: There exists an economic leader region-1 and economic following region-2.

Assumption 2: Technical innovation brings about new product functions, or innovative products increase the difference in and diversity of the product.

Assumption 3: Technical innovation increases the number of different products, N.

Assumption 4: Production of a different commodity requires only the input of labor force and every unit of production occupies a unit of labor force.

Assumption 5: Region-1 needs to invest in $\frac{a}{K_n}$ units of labor to develop new products where a represents productivity of the deployment while technology stocks is given by $K_n = f(n)$; Region-2 is in constant innovation and its manufacturers input $\frac{a_m}{K_m}$ units of labor to imitate the technology where a_m is the productivity of imitation, K_m is the current stock of technical knowledge. When region-2 absorbs, digests, and reinvents foreign technology, K_m will increase with the accumulation of experience, but will continue to rely on region-1 to produce N_1 different products.

Assumption 6: When region-1 can produce enough innovative products, region-2 can easily imitate their products and also reduce their cost. So $K_m = K_m\ (N_1, N_2)$ is assumed to be a homogeneous linear, increasing function of N_1 and N_2. N_2 represents the number of products that could be used by region-2 and $N_2 \le N_1$.

Assumption 7: N_1^* represents the number of innovative products in region-1, while N_2^* represents the number of imitation products in region-2.

3.3.2 Imitation Model of Technical Innovation in Two Regions

3.3.2.1 Market Clearing Conditions in Two Regions

According to the above assumptions, $x_{i(i=1,2)}$ represents the sale volume of the products in the two regions, that is

$$x_i = \frac{(p_i)^{-\varepsilon}}{N_2(P_2)^{1-\varepsilon} + N_1(P_1)^{1-\varepsilon}} \quad (i = 1,2) \tag{3.9}$$

where

p_1 and p_2 represent the prices of different products in region-1 and region-2
$\varepsilon = \frac{1}{1-\alpha}$ the substitution elasticity of the two products
α the preference coefficient of different products with $0 < \alpha < 1$
$i = 1,2$ i_1 and i_2 separately represent region-1 and region-2

Manufacturers in region-1 have a demand for labor force, N_1x_1, while the research department requires $\frac{\alpha N}{K_n}$ for labor. The balanced demand in the labor market is as follows:

$$\frac{\alpha N_1^*}{K_n} + N_1 x_1 = L_1 \tag{3.10}$$

where

L_1 represents total supply of labor force in region-1

N_1^* represents the number of innovative products in the same region

Similarly, N_2x_2 represents the manufacturers' demand for labor in region-2, where the labor force input in the imitation activity is $\frac{\alpha_m N_2^*}{K_m}$. Clearing the labor market therefore yields

$$\frac{\alpha_m N_2^*}{K_m} + N_2 x_2 = L_2 \tag{3.11}$$

where

L_2 represents total labor force supply in region-2

N_2^* is the number of imitation products in this region

Assuming that the labor allocation among industries in two regions stays fixed and $\xi_i = \frac{N_i}{N}$ $(i=1,2)$ is the share of region-i in all the different product categories. In the long-term, ξ_1, ξ_2 approach a constant value, and the growth rate g of the production categories in every area converges, that is $g_1 = g_2(g_i = \frac{N_i^*}{N_i}$, $i = 1,2)$. For $N = N_1 + N_2$, $g = \xi_1 g_1 + \xi_2 g_2$. In a steady situation, $g = g_1 = g_2$. The imitation rate $m \equiv \frac{N_2^*}{N_1}$ represents the rate of products being imitated per unit of time in region-1. Accordingly $m = \frac{g_2 \xi_2}{1-\xi_2}$; when $g_2 = g$, then $\xi_2 = \frac{m}{g+m}$.

As a result, compared with the rate of innovation g, the higher the rate of imitation m, the higher the product proportion ξ_2 will be in region-2 in the long-term.

3.3.2.2 Relation between the Imitation Rate and Innovation Rate

If manufacturers in region-1 have developed technology that has not been imitated, it can be assumed that it could get a profit $\pi_1 dt$ in time dt. The companies face the risk of their products being imitated by manufacturers in region-2; it can therefore be assumed that there are $N_2^* dt$ dt kinds of products being imitated in time dt. Then the possibility that manufactures in region-1 lose the monopoly status in that time interval is $\frac{N_2^* dt}{N_1}$. Assuming that the capital lost is v_1, the profit is $\dot{v}_1 dt$. Then the total expected income of manufacturers in region-1 is as follows:

$$\pi_1 dt - \frac{N_2^* \, dt}{N_1} v_1 + \left(1 - \frac{N_2^* \, dt}{N_1}\right) \dot{v}_1 dt \tag{3.12}$$

If the loan is of size v_1, dividing by $v_1 dt$, and letting $dt \to 0$

$$\frac{\pi_1}{v_1} + \frac{\dot{v}_1}{v_1} - \frac{N_2^*}{N_1} = r_1 \tag{3.13}$$

where r_1 represents the bond profit of manufactures in region-1. $r_1 = \rho$ in the long-term, and ρ is the discount rate. That is

$$\frac{\pi_1}{v_1} = \rho + g + m \tag{3.14}$$

$$\pi_1 = \frac{(1-\alpha)w_1}{\alpha(1-\xi)N}(L_1 - ag) \tag{3.15}$$

where w_1 represents the marginal cost of the manufacturers in region-1.

By a combination of $\xi_2 = \frac{m}{g+m}$, and Formulas 3.13, 3.14, and 3.10, the relationship between the hidden long-term rate of innovation g and imitation rate m when the market is clear in region-1 and $g>0$, can be obtained as shown here in Formula 3.16:

$$\frac{1-\alpha}{\alpha}\left(\frac{L_1}{a} - g\right)\frac{g+m}{g} = \rho + g + m \tag{3.16}$$

Similarly, the total value for the manufacturers in region-2 also remains unchanged.

$$\pi_2 = \frac{(1-\alpha)w_2}{\alpha N_2}(L_2 - a_m g) \tag{3.17}$$

where w_2 presents the marginal cost of the manufacturers in region-2.

With different rates of innovation in the stable state, the greater g is, the higher the capital losses of manufactures in region-1, which shows that the right side of Formula 3.16 will be larger. This signifies that the actual capital cost of the country's manufacturers in a steady state is higher. At the same time, in equilibrium where the value of g is large, the profit of the manufacturers in region-1 is lower as Formula 3.15 indicates. The reason for this situation is that on one hand, the high rate of innovation implies that the R&D sector employs more workers and leaves the smaller labor force to work for manufactures, on the other, region-1 has a higher quotient in the total number of products in the stable state, implying ξ_2 is smaller $\left(\xi_2 = \frac{m}{g+m}\right)$ and the distribution of the labor force will become even more sparse.

In the stable state with different imitation rates, the greater the rate of imitation m, the higher the risk of manufacturers being driven out of the market in region-1, and the higher the realistic cost of capital for them. As an increase in the imitation rate m can improve the profit rate of the manufacturers in region-1 the rate of imitation becomes higher. Because $\xi_2 = \frac{m}{g+m}$, the less the product share of region-1, the bigger its sale volume for a certain number of products. Imitation has more impact on the profits than on the cost of capital for the same flexibility of demand, so the high rate of innovation matches the high rate of imitation to ensure the validity of Equation 3.16.

3.4 Measurement of Technology Transfer

The measure and analysis of the "strength" generated by regional technology transfer and flow are the necessary premise for achieving reasonable control on technology transfer and its effect. It not only reveals the length, depth, and breadth of economic and technological relevance in a certain time period in different regions, but also reflects the influence of economic development in different regions on its strategic, superior, and general industries.

3.4.1 Model of Technology Transfer

In physics, $F = G\frac{m_1 m_2}{r^2}$ describes the attraction between two objects and depicts attraction between two objects with a mass of m_1, m_2 when the distance between them is r; the greater the mass of the objects and the closer they are to each other, the greater the attraction will be. There are some similarities between the technology transfer between two regions and the gravitational attraction between two objects, namely, the greater the interest difference, the stronger the economic strength and the closer they are located, the stronger the trend of technology transfer will be. As a result, the ideology and measurements of gravity in physics can be applied here. Assuming that there are two areas i and j, whose technology levels are r_i and r_j respectively, the transfer gravity of technology elements is given by

$$
\begin{cases}
F_m^{ij} = k_m^{ij} \dfrac{G_m^i G_m^j}{S^q} (p_m^i a_m^i - p_m^j a_m^j) & p_m^i a_m^i > p_m^j a_m^j \text{且} r_m^i > r_m^j \\
0 & p_m^i a_m^i < p_m^j a_m^j \text{或} r_m^i < r_m^j
\end{cases}
\quad (3.18)
$$

F_m^{ij} represents the gravity between the two regions *i* and *j* in industry *m*. Technical elements will flow from the region with higher benefit returns to the other one. It will also flow from the region at a higher technical level and with higher productivity elements to the other one:

r_m^i and r_m^j are the technical levels of regions *i* and *j* respectively in industry *m*.

k_m^{ij} represents the coefficient of technology transfer between regions *i* and *j* in industry *m*, and is related mainly to policy factors. As policies are easy to transfer, technical elements can be easily transferred. The greater the value of k_{ij}, the greater the transfer capacity.

p_m^i and p_m^j represent influence coefficients of industrial development for regions *i* and *j* respectively in industry *m*, and are related mainly to the comprehensive matching of factors that affect the development of technical elements, such as living environment, geography, openness, communication, language, and cultural affinity. Regions with better traffic conditions, more comfortable living environment, similarity in language and culture, and greater affinity will have greater attraction to technical elements, and the value of p_m^i and p_m^j will be greater.

G_m^i and G_m^j represent the GDPs (Gross Domestic Product) of regions *i* and *j* respectively in industry *m*. The better developed the industry, the more the output will be, and the more the attraction to similar industries.

a_m^i and a_m^j represent return rate of transfer of the regions *i* and *j* respectively in industry *m*.

S is the distance between the regions *i* and *j*, which reflects the influence of distance on technology elements flow. The less the distance between two regions, the faster the technology transfer, the more the attraction and vice versa.

Q stands for the index of the distance, and it reflects the influence of geographical conditions, like transportation and geographic barriers to technology elements transfer, With improvement in transportation conditions, green time will be shortened, the transportation will be more convenient, *q* will be smaller, and the negative influence will also be smaller, which will, in turn, result in higher attraction.

3.4.2 Empirical Study on Technology Transfer in China

Technique is the result of a comprehensive transfer of technology elements, and it is the sum of different technology production factors. In this study, based on the geographical distribution and city scale, certain electronic and communications equipment manufacturing units in some representative provinces and cities in China

(Anhui, Fujian, Shanghai, Hangzhou, Shanxi, Chongqing, Qingdao, Harbin, Nanjing) are selected, and the capability of technology transfer from 2001 to 2003 is estimated.

3.4.2.1 Data Specification

1. The selection of industry
   m represents high-tech industries like electronics and communications equipment manufacturing industry.
2. Coefficients of transfer technique
   On the basis of the Chinese regional policies to attract foreign investment, the following assumptions are made: when technology elements flow to eastern regions, $k_m^{ij} = 1$, when they flow to the central regions, $k_m^{ij} = 0.9$, and when they flow to the west, $k_m^{ij} = 0.8$; (special economic zone is classified as East region).
3. Influence coefficients of industrial development
   For mechanism analysis of technology transfer, gray clustering analysis is used; the whitening weight function is generated and the importance of different indicators is distinguished by using Delphi basic language, conference consulting, and comparing step by step. The cities under calculation were divided into three categories A, B, and C and their values are 0.9, 0.8, and 0.6 respectively. The results are shown in Figures 3.1 and 3.2.
4. Return rate of industrial transfer
   a_m^i and a_m^j represent the comprehensive indexes of industrial economic benefit. This new index evaluation system was introduced in China in 1998, and is composed of seven indicators, which are calculated as follows: each reported practical value of the seven single indicators of industrial economic benefit is divided by its national standard value and multiplied by its own weight, the sum of the above results is then divided by the sum of the weights.

   IA special relative number gives the comprehensive industrial economic benefit calculated from the following seven aspects: profitability, development capacity, operational risks, speed of reproduction cycle, economic benefit derived by reducing the cost, production efficiency, and the cohesion between production and marketing. So it reflects the gross index in the operation quality of the industrial economy, and the actual level and development trend of industrial economic benefit in different industries as well as different regions can be checked and evaluated at the same time. These numbers reflect the panorama of the economic situation. Therefore transfer return rate is chosen as the indicator in this study.

   The index of industrial economic benefit is composed of seven indicators: the total assets contribution ratio, hedging and proliferation ratio, asset-liability ratio, asset turnover ratio, profit margin of the cost, all labor productivity, and product sales rate.

Table 3.1 The Classification of National Technical Development Influence Coefficients

Indicators		*Unit*	*Weight*	*Weaker Category*	*General Category*	*More Categories*
Security honesty	The proportion of tertiary education staff	Percent	7	$5 \le x_1^1 < 15$	$15 \le x_1^2 < 45$	$45 \le x_1^3 < 60$
	The employment rate	Percent	10	$0.4 \le x_2^1 < 0.5$	$0.5 \le x_2^2 < 0.65$	$0.65 \le x_2^3 < 0.7$
	Criminal cases per million people	People	5	$50 \le x_3^1 < 100$	$100 \le x_3^2 < 200$	$200 \le x_3^3 < 250$
Environmental science and technology	The number of scientific research, development, and intermediary departments	One	5	$200 \le x_4^1 < 600$	$600 \le x_4^2 < 2{,}000$	$2{,}000 \le x_4^3 < 3{,}000$
	R&D financing/GDP	Percent	7	$0.5 \le x_5^1 < 0.8$	$0.8 \le x_5^2 < 1.6$	$1.6 \le x_5^3 < 2.0$
	Technology market turnover	Billion Yuan	5	$0.5 \le x_6^1 < 10$	$10 \le x_6^2 < 100$	$100 \le x_6^3 < 150$
Traffic and geography	Passenger turnover	Billion km	5	$220 \le x_7^1 < 400$	$400 \le x_7^2 < 800$	$800 \le x_7^3 < 1{,}000$
	Goods turnover	Billion km	5	$150 \le x_8^1 < 200$	$200 \le x_8^2 < 400$	$400 \le x_8^3 < 450$
	Passenger traffic	Billion people	5	$3 \le x_9^1 < 8$	$8 \le x_9^2 < 15$	$15 \le x_9^3 < 20$
	Cargo traffic	Billion ton	5	$5 \le x_{10}^1 < 8$	$8 \le x_{10}^2 < 12$	$12 \le x_{10}^3 < 15$

Index		Unit	Weight	Weaker Category	General Category	Strong Category
City life	Per ten thousands people, public transport vehicles	One	7	$1 \le x_{11}^1 < 4$	$4 \le x_{11}^2 < 9$	$9 \le x_{11}^3 < 12$
	Urban public green area per capita	Yuan	7	$2 \le x_{12}^1 < 5$	$5 \le x_{12}^2 < 8$	$8 \le x_{12}^3 < 12$
	Health agencies	One	6	$1 \le x_{13}^1 < 9$	$9 \le x_{13}^2 < 20$	$20 \le x_{13}^3 < 28$
	Per capita living space for urban residents	M^2	7	$15 \le x_{14}^1 < 20$	$20 \le x_{14}^2 < 27$	$27 \le x_{14}^3 < 30$
Economic environment	The total investment in fixed assets	Billion Yuan	8	$600 \le x_{15}^1 < 2{,}000$	$2{,}000 \le x_{15}^2 < 7{,}000$	$7{,}000 \le x_{15}^3 < 10{,}000$
	Per annual consumption	Yuan	5	$5{,}000 \le x_{16}^1 < 6{,}500$	$6{,}500 \le x_{16}^2 < 8{,}500$	$8{,}500 \le x_{16}^3 < 10{,}000$
	Per capita GDP	Yuan	6	$5{,}500 \le x_{17}^1 < 8{,}000$	$8{,}000 \le x_{17}^2 < 12{,}000$	$12{,}000 \le x_{17}^3 < 18{,}000$

Source: Data from Industrial Statistics Yearbook of provinces and cities (2001–2003) and the China Statistical Yearbook (2001–2003).

Table 3.2 Technique Development Influence Coefficients for the Communications Equipment Manufacturing Industry in Some Provinces and Cities in China

Region	*Time*	p_m^i	*Time*	*Region*	p_m^i	*Region*	*Time*	p_m^i
Anhui	2001	0.6	Hangzhou	2001	0.9	Qingdao	2001	0.8
	2002	0.6		2002	0.9		2002	0.8
	2003	0.6		2003	0.9		2003	0.9
Fujian	2001	0.6	Shanxi	2001	0.6	Haerbin	2001	0.8
	2002	0.6		2002	0.6		2002	0.8
	2003	0.8		2003	0.6		2003	0.9
Shanghai	2001	0.9	Chongqing	2001	0.6	Nanjing	2001	0.8
	2002	0.9		2002	0.6		2002	0.9
	2003	0.9		2003	0.6		2003	0.9

$$\text{The total assests of the contribution ratio (percent)} = \frac{\text{Total profits} + \text{Total taxes} + \text{Interest expense}}{\text{Total assets}} \times \frac{12}{\text{the cumulative number of months}} \times 100\,\text{percent}\,(\text{Standard value } 10.7, \text{ weights } 20) \tag{3.19}$$

$$\text{The hedging and proliferating ratio (percent)} = \frac{\text{The owner's rights at the end of the reporting period}}{\text{The owner's rights last year}} \times 100\,\text{percent}\;(\text{Standard value } 120, \text{ weights } 16) \tag{3.20}$$

$$\text{The asset-liability ratio (percent)} = \frac{\text{Total liabilities}}{\text{Total asset}} \times 100\,\text{percent}\;(\text{Standard value } 60, \text{ weights } 12) \tag{3.21}$$

$$\text{The asset turnover ratio (percent)} = \frac{\text{Product sales revenue}}{\text{Average balance of total current assets}} \times \frac{12}{\text{the cumulative number of months}}\,(\text{Standard value } 1.52, \text{ weights } 15) \tag{3.22}$$

The profit margin of the cost (percent)

$$= \frac{\text{The total profits}}{\text{The total cost}} \times 100 \text{ percent (Standard value 3.71, weights 14)} \quad (3.23)$$

The all labor productivity (percent)

$$= \frac{\text{The industrial added value}}{\text{The average number of employees}} \times \frac{12}{\text{the cumulative number of months}} \text{(Standard value 16,500, weights 10)} \quad (3.24)$$

The industrial products sales ratio (percent)

$$= \frac{\text{The sales of industrial output value}}{\text{The industrial output value at current prices}} \times 100 \text{ percent (Standard value 96, weights 13)} \quad (3.25)$$

When the asset-liability ratio exceeds 60 percent, the score of the rate

$$= \frac{\text{Index value-not allowed value 100}}{\text{60-Not allowed value 100}} \times 12 \quad (3.26)$$

The comprehensive index of industry economic benefit
= Σ [(reporting value of the indicator ÷ national standard value) × weights] ÷ total weights

5. GDP of industry
   G_m^i and a_m^j represent the total output value in the electronic and communication equipment manufacturing industry of different regions.
6. The distance between two cities
   S stands for the economic distance between two cities.

   Here, Gaoruxi's approach is adopted to calculate the economic distance, which is based on space distance. After two stages of corrections, the economic distance is obtained as follows:

$$E = \alpha\beta D \quad (3.27)$$

where

E represents the economic distance
D represents the space distance
α and β represent correction weights

Table 3.3 Correction Weights for Commuting Distance and Economic Gap

Commuting distance correction weight, α							
Combination of vehicles weight α	Train 1	Car 1.2	Ship 1.5	Train and car 0.7	Train ship 0.8	Car and ship 1.1	Train, car, and ships 0.5
The economic gap correction weight, β							
Per capita GDP of surrounding cities/per capita GDP of core cities				>70 percent	70 percent ≥ rate ≥ 45 percent		<45 percent
Weight β				0.8	1.0		1.2

α is the correction weight in the first correction (the correction weight of commuting distance), and its value is determined by city transportation conditions and β is the correction weight in the second correction t (the correction weight of the economic gap), and its value is determined by dividing the per capita GDP of the surrounding urban area by that of the core cities. Specific values are shown in Table 3.3.

After calculating the economic distance, the other cities are divided into four groups from short to long according to the economic distance; the greater the economic distance, the larger the team number. Finally the valued team number to discriminate the value E of economic distance in the city is evaluated.

7. Distance index

   Q represents the actual traffic level calculated by comprehensive measurements of the region. To simplify the calculation, q is taken as 1.

3.4.2.2 Results of Calculations

According to the characteristics of China's economic development and geographical location, by using Formula 3.17, we could calculate the transfer capacity of technique of its communications equipment manufacturing industry. The results are as shown in Table 3.4.

3.4.2.3 Analysis of Results

From the above calculation, it is clear that:

1. After multiplying the comprehensive indicator of industry benefit by the transfer coefficient of production elements, the value of the two regions can be obtained; the difference between these values affects the capacity of technique transfer most. It can be concluded that transfer returns and development environment are the main factors which affect productivity transfer.
2. In addition, the productivity transfer is affected by the industrial GDP, and the higher the degree of industrial agglomeration (higher the GDP), the more the attraction of the region to the productivity of other regions; in other words, productivity influences productivity transfer in China.
3. For some underdeveloped regions like Shanxi and Anhui, besides accelerating the development of their own industries, they have to improve on optimizing the development environment, both in the software aspect (policy system) and hardware aspect (infrastructure) so as to attract technique.
4. The impact of distance on technology transfer shows a falling trend in successive years, which, to some extent, shows that with improved traffic conditions in China, the effect of distance on productivity transfer is getting smaller and smaller.

Table 3.4 The Transfer Capacity of Technique for the Communications Equipment Manufacturing Industry in Some Provinces and Cities of China

Source / Region Destination	Year	Anhui	Shanghai	Shanxi	Fujian	Chongqing
Anhui	2001		–128.9	0.1	–8.8	0.0
	2002		–84.6	0.1	–12.6	0.0
	2003		–224.1	0.1	–56.3	0.1
Shanghai	2001	128.9		3.6	284.8	3.4
	2002	84.6		3.5	119.5	2.6
	2003	224.1		6.8	–716.4	9.0
Shanxi	2001	–0.1	–3.6		–0.2	0.0
	2002	–0.1	–3.5		–0.5	0.0
	2003	–0.1	–6.8		–1.1	0.0
Fujian	2001	8.8	–284.8	0.2		0.9
	2002	12.6	–119.5	0.5		0.8
	2003	56.3	716.4	1.1		5.8
Chongqing	2001	0.0	–3.4	0.0	–0.9	
	2002	0.0	–1.5	0.0	–0.8	
	2003	–0.1	–7.3	0.0	–5.8	
Nanjing	2001	170.0	–673.5	1.4	30.8	0.5
	2002	261.8	686.0	2.4	93.6	0.5
	2003	251.4	–547.3	2.1	–167.0	1.0
Qingdao	2001	2.3	–110.1	0.5	–2.1	0.1
	2002	8.1	28.6	1.5	9.6	0.1
	2003	6.7	–193.8	1.3	–45.6	0.2
Haerbin	2001	0.0	–1.1	0.0	–0.1	0.0
	2002	0.0	–0.8	0.0	–0.1	0.0
	2003	0.0	–0.3	0.0	0.0	0.0

Table 3.4 (continued) The Transfer Capacity of Technique for the Communications Equipment Manufacturing Industry in Some Provinces and Cities of China

Source / Region / Destination	*Year*	*Anhui*	*Shanghai*	*Shanxi*	*Fujian*	*Chongqing*
Hangzhou	2001	12.8	–1677.5	0.3	–10.3	0.3
	2002	56.4	2229.4	6.9	220.7	0.8
	2003	98.5	4718.5	36.4	82.3	2.4
	Nanjing	*Qingdao*	*Haerbin*	*Hangzhou*	*Total*	
Anhui	–170.0	–2.3	0.0	–12.8	–322.7	
	–261.8	–8.1	0.0	–56.4	–423.4	
	–251.4	–6.7	0.0	–98.5	–636.8	
Shanghai	673.5	110.1	1.1	1677.5	2883.0	
	–686.0	–28.6	0.8	–2229.4	–2732.9	
	547.3	193.8	0.3	–4718.5	–4453.5	
Shanxi	–1.4	–0.5	0.0	–0.3	–6.1	
	–2.4	–1.5	0.0	–6.9	–14.9	
	–2.1	–1.3	0.0	–36.4	–47.9	
Fujian	–30.8	2.1	0.1	10.3	–293.2	
	–93.6	–9.6	0.1	–220.7	–429.3	
	167.0	45.6	0.0	–82.3	910.1	
Chongqing	–0.5	–0.1	0.0	–0.3	–5.2	
	–0.5	–0.1	0.0	–0.8	–3.8	
	–1.0	–0.2	0.0	–2.4	–16.8	
Nanjing		24.5	0.2	126.9	–319.2	
		19.6	0.4	–196.3	867.9	
		37.7	0.0	–625.0	–1047.0	

(continued)

Table 3.4 (continued) The Transfer Capacity of Technique for the Communications Equipment Manufacturing Industry in Some Provinces and Cities of China

Source Region / *Destination*	*Nanjing*	*Qingdao*	*Haerbin*	*Hangzhou*	*Total*
Qingdao	−24.5		0.0	−0.6	−134.5
	−19.6		0.1	−9.8	18.5
	−37.7		0.0	−30.5	−299.5
Haerbin	−0.2	0.0		−0.1	−1.5
	−0.4	−0.1		−0.3	−1.7
	0.0	0.0		−0.1	−0.4
Hangzhou	−126.9	0.6	0.1		−1800.5
	196.3	9.8	0.3		2720.7
	625.0	30.5	0.1		5593.6

5. Positive and negative transfer capacity represent the direction of productivity transfer; positive values show that there is the attraction, while the negative value indicates that there is a reverse transfer capacity.
6. Technology transfer capacity is consistent with the increasing rate of local industrial GDP. With the increase in gross industrial output value, productivity should improve and vice versa, which is consistent with the actual situation.

Technology transfer and consequent reorganization is inevitable with technology transfer. It is the inherent requirement for technological development. With the rapid development of technology and sustained openness of the entire economic system, various technology elements find superior application, by means of transverse and vertical movements and reorganization in different kinds of patterns. In the process of technology transfer and reorganization, the speed, scale, structure, nature, and the results differ with the differences in economic conditions, absorptive capability of transfer mechanisms, and so on.

Chapter 4

Cost and Benefit of Technology Transfer

Technology transfer has become the primary driving force behind the development of economy and the foundation of sustainable development in human society. The acquiring of profits plays a vital and basic role in technology transfer; the cost has to be paid essentially to acquire such profits. Cost is essential to reap benefits to some extent and obtain beneficial opportunities regardless of whether such interest is predictable and whether it is direct. Under certain kinds of market and system arrangements, a mutually promotional cycle can take shape if satisfactory benefit can be derived at some cost by optimal allocation of technical resources to proper flows of techniques, which speeds up the development of social economy. Otherwise, such anticipated goals can hardly be reached. It is an understanding of the cost and benefit of technology transfer in a scientific way that can have a profound and practical impact on the rational transfer of technology.

4.1 Cost of Technology Transfer

4.1.1 Connotation of Cost

The concept of cost is considered fundamental in economics. As objects and aspects studied may differ, the concept of cost varies greatly in different fields of study. In management accounting, cost is defined as resource that has to be sacrificed or abandoned in the pursuit of a fixed goal; in financial accounting science, cost is the price of obtaining assets. According to the principle of Marx's political economics, cost, being an important part of the commodity value, is considered the expense for labor, material resources, and finance in economic activities with the purpose of gaining certain goods. The essence of cost is the sum of transfer values consumed in the process of production, in materialized and living labor.

In economics, the category of cost has been expanding radically with the emergence of management accounting in the 1920s, giving birth to various concepts such as project cost, responsibility cost, quality cost, capital cost, opportunity cost, sunk cost, and variable cost as well as constant cost. Since the 1950s, economics and society have been both differentiated and highly integrated, while the concept of cost has been generalized. On one hand, the economic concept of cost was introduced into many subjects and absorbed into them, on the other hand some useful ideas of cost from various subjects were absorbed into economics in return. As a result, a complicated system of these concepts was formed with the scope of cost enriched gradually and its extension expanded continuously. Therefore, the concept of cost cannot be rigidly defined scientifically as it is universal to every subject.

In general, cost is the consumption of resources, which can be denoted as money. In spite of this denotation, cost is by no means represented only by money and expenses. Various kinds of expenses in production and management activities can be considered as cost in the narrow sense along with the concept of "profit." The concept of cost in the narrow sense can be measured in many ways for example in RMB100, ten computers, etc. This can also be represented in currency as a finite amount.

In addition, there are costs that cannot be measured, such as the panic that results from unemployment and social turbulence because of inflation. All these are part of the various expenses for a certain economic entity (including economic cooperation and natural person) with specific economic activities. These kinds of cost include not only actual expenses and loss already suffered, but also various prices to be paid in anticipation. The category of cost in the general sense comprises social cost, system reform cost, reform cost, opportunity cost, and transaction cost in the contemporary study of microeconomics.

4.1.2 Factors That Affect the Cost of Technology Transfer

4.1.2.1 Asymmetric Information

It should be clear that technology transfer is often implemented as asymmetric information, which is a key factor in the technology transfer cost.

Asymmetric information denotes the information that is with the transferors (generally those who transfer technology out) and not with the transferees (those to whom technology is transferred). It is likely that the owners of information may deceive those who are ignorant or lack the information with their information superiority. In the process of technology transfer, the technology transferors who have information superiority have almost all the factors of technology transfer, while the technology transferees know little about the technology to be transferred, placing themselves in the inferior position of obtaining information. In this case, information asymmetry occurs. Several constraints, attributable to the existence of such asymmetry for a long period of time, are confronted in the flow of information:

1. Existing technology level constraint
   It is quite common that the transferees of technology do not understand the core of the transferred technology thoroughly owing to the limit in their existing technology, talents, and levels of management.
2. Time-delay constraint
   On one hand, technology transferors are reluctant to transfer (transit) the newly developed technology at once; on the other hand, it takes time for the transferees to absorb key points of technology completely.
3. Constraint of technology transferors' own interest
   The value of technology is comparatively high, especially its privacy. On some occasions the transferors may deceive the transferees by their information superiority, by revealing only key points partly or exaggerating the effect of the technology.
4. Constraint of information cost
   Cost is essential for collecting information, and increase in the cost makes the flow of technology impossible. Under such a constraint, transferees need to be fully aware of the technology transferred.

   The existence of these objective and subjective constraints may increase the risk of technology transfer.

4.1.2.2 Financial Benefit

Financial benefit is the most basic and important aspect of technology transfer, which affects transfer cost. Strictly speaking, economic cost is a concept with wide connotation in economics, including direct and indirect financial benefits; calculating it is a complicated task. Therefore, the financial benefit of technology transfer can be simplified to the calculation of profits with taxation taken into account, being the most basic and important part after achieving technological progress.

The financial benefit can be determined on the basis of the levels of technology, quality, market competitiveness, and the ability to absorb technology. It can be also determined on the basis of the amount of investment and investment recovery period for the new tech applications and new goods investment as well as some recent economic and social environment factors.

The predictions on financial benefit may vary greatly according to the different views of both sides involved in the technology transfer, which requires negotiation from both sides with an objective attitude or even assistance of a third party, for example, evaluation by asset appraisal firms.

4.1.2.3 Supply and Demand

This factor includes two aspects, namely the supply–demand relation in the technology market and the materialization of such a relation in the commodity market. If the technology belongs to the buyer market, the technology transfer price is likely to reach the price expected by the buyer, and vice versa. For those who receive the technology transfer, their bidding for it and the price they offer depend mainly on the material products related to this technology, although the value in related technology market will be based on the supply–demand relation. If certain kinds of goods are in great demand, the technology closely related to them is bound to become popular, resulting in the rise of price. Similarly, if supply of certain goods exceeds their demand in the technology market, the prices of these goods may go down or even touch a low. Therefore, the fluctuation of prices has a great impact on the variation of cost, while the cost of technology transfer is affected by supply and demand.

4.1.3 Classification of Technology Transfer Cost

Although, it is harder to define technology transfer cost compared to the similar attempt to define cost, great effort has been made in this book to analyze technology transfer cost on different aspects.

4.1.3.1 Direction of Technology Transfer

The direction of technology transfer can be divided into technology transferred in and transferred out. Consequently, its cost can be naturally classified as transfer-in cost and transfer-out cost.

1. Transfer-in cost
   Transfer-in cost includes
   - Technology searching cost is the cost incurred in the search for developed technology to replace the technology that is immature or lacks elements that the owner is reluctant to research by himself, for example, the cost incurred for purchasing technical intelligence and outdoor exploration.
   - Technology importing cost includes cost of purchasing advanced technology from technology transferors, importing it through communication channels such as academic conferences, exhibitions, and visits, and training personnel to use the new technology.
   - Cost of technology loss refers to the cost of training when the trainees fail to acquire or absorb new technology.
2. Transfer-out cost
   Transfer-out cost includes
   - Technology transition cost refers to the cost incurred in the process of transferring part or all of the technology obtained by transferors to others to gain more profits.

- Market segmentation cost arises when the transferors and transferees compete in the development of technology. As a result there is a division of the market for the transfer of technology, which affects the amount of profit although the leading position of the transferors stays unchanged.
- Talent supplement cost is incurred when new talent is sought, out of a fear of greater loss, to fill the vacancy created by some talented people who leave. In such a situation, the cost may increase when new suitable talent is acquired through headhunting companies and the mass media.
- Talent attracting cost is incurred when newly recruited talent follows a trend of leaving soon after joining. To retain and attract fresh talent, some incentives such as pay rise and increased welfare measures will lead to increase in cost.
- Cost of information loss is the result of people who leave their organizations taking away their project and technology knowledge acquired in the organization, as the passing of such knowledge to other organizations may lead to a great loss.

4.1.3.2 Factors of Technology Transfer

Technology transfer cost can be divided into the cost of technology talent transfer, cost of technology transfer, and cost of technology carriers.

The cost of technology talent transfer refers to all the expenses involved from the beginning of the talent transfer (voluntary and involuntary) to the actual substitution by companies, including the tangible cost which occurs when new talent is acquired (for example, the expenses of advertising, interview, and training, etc.), and the intangible cost which has little or no direct link with actual expenditure (for example, management expenses, decrease in labor efficiency, etc.).

The technology transfer cost can be considered from two points of view—of the transferors and the transferees. In the view of the transferors, their cost mainly comprises the cost of developing new technology plus the cost of this failure, the cost of the technology transfer, and the cost of training, guidance, and maintenance after the transfer. It should also consider the loss of interest due to the division of market share after technology transfer if such transfer takes place in some related companies. In the view of the transferees, the cost includes the transaction prices while transferring, opportunity cost brought about by the transfer, and expenses during the process of circulation.

4.1.3.3 Willingness in Technology Transfer

From the point of view of whether technology transferors are willing to make the transfer, technology transfer cost can be classified as voluntary cost and involuntary cost.

Voluntary cost refers to the cost resulting from the decline of profits because of expenses for propaganda on the transferors' advanced technology with the intention to improve sales and influence. This leads to the loss of their monopoly status after others' imitate their technology, resulting in the segmentation of their market share. In general, it is important that transferors ensure that the anticipated benefit of such transfer will outweigh this cost considerably.

Involuntary cost refers to the cost resulting from the loss of market share and leading status of the transferors who monopolize the advanced technology market and are reluctant to transfer but are obliged to release or transfer part or all of the technology in consideration of the overall interest, limits of policies, and other irresistible factors. It also includes the cost resulting from the loss of market share because their talents and technology are taken by others.

4.1.3.4 Nature of Technology Transfer

Two kinds of benefits namely explicit benefit and implicit benefit occur in the transfer of technology. Similarly, technology transfer cost can be divided into explicit transfer cost and implicit transfer cost.

Explicit transfer cost includes expenses paid in the process of transfer, such as the expenses incurred in the search for the right talent and relevant technology, the expenses of training and transactions, etc.

Implicit transfer cost includes the loss of customers, change of environment, decline in organization efficiency, increase in the number of competitors, the loss of commercial opportunity, etc.

4.1.3.5 Participants of Technology Transfer

Considering the participants of technology transfer, technology transfer mainly involves enterprises (or industries) and governments; the cost can be classified as enterprise (or industry) transfer cost and government transfer cost.

In the mechanism of the market, technology transfer is driven by economics. Technology transfer is, in some sense, a kind of reaction to the search for better interest opportunities and bigger economic space when part of the technology works in the market mechanism. Companies in search of interest and domination in technology transfer must pay the cost for their transfer, including transaction cost, opportunity cost, circulation cost, etc.

The entire local society is keenly interested in technology transfer, which brings about indirect benefit especially social benefit to it. This kind of benefit, which is beyond the anticipation of enterprises (industries), is obtained when governments pay the cost. Government cost includes actual expenses that propel technology transfer and expenditure used for several kinds of policies. In addition, it covers all kinds of implicit cost in economy and ecology used for government behavior and decision-making.

4.1.4 Nature of Technology Transfer

4.1.4.1 Typical Irreversible Investments Partly Belonging to Capital Cost

It can be demonstrated that technology transfer cost partly belongs to capital cost in view of the investments on various kinds of equipments and facilities in the production system. The capital feature of technology transfer cost can be better described as the acquisition of knowledge or skills. For instance, to master a kind of foreign language, individuals should spend money on textbooks, hiring tutors and some initial investments which are irreversible such as time and effort spent in the study of the language. Although such investment in the study can be transferred to others in some forms, these investments will still be considered as personal investments that cannot be transferred out totally.

4.1.4.2 Cost Variation in Different Domains and Aspects

Obtaining technology in an uncertain domain is more expensive than in a comparatively familiar domain. People with the same experience or in the same industries, as cited by Adam Smith, tend to exchange information with each other more easily and effectively compared to those without the same experience or in different industries. People with the same experience basically have the same information cost since they share the direction of information investment, whereas those with different experiences have different costs in obtaining information.

4.1.4.3 Technology Transfer Cost and Nonlinear Variation of Benefit

There are certain links where nonlinear between-technology transfer cost and technology transfer benefit. Apart from the technology cost, the benefit of technology transfer is influenced by many factors in the process of production. Moreover, technology transfer brings about not only expected benefit for transferees but also the promotion of technology in the entire industry and local areas.

4.1.4.4 Nature of Shift in Technology Transfer Cost

Many types of technology, where the cost is borne by all the citizens, are public in nature. However, the products and services of technology shared by the same tax payers may differ, for example, some of them who do not share the technology have to pay the tax whereas some others who share the technology pay little or even nothing. It is difficult to calculate technology transfer cost exactly owing to the wide outreach of its concepts and its combination with other concepts.

4.2 Benefits of Technology Transfer

4.2.1 Connotation of Benefit

The word "benefit" denotes effect and interest in Chinese. Effect is the result (usually good result) of certain power, actions, and factors, while interest refers to certain profits which can be divided into material interests and spiritual interests. In general, the concept of interest is related to increase, profit, and development.

The concept of benefit has always been controversial. There is a point of view accepted by the majority that benefit refers to the comparison between input (or cost) and output (or gain). Even though accepted by the majority, this is far from accurate. As a matter of fact, the interest can be interpreted in several ways, an integration of meanings at different levels. The following definitions are some interpretations:

1. In general, benefit refers to the comparison between input (cost) and output (gain). The benefit is good when less is spent on input and much is gained as output; and vice versa.
2. The concept of benefit is relative and can be measured by quality and quantity. The quality measure refers to the comparison of the quality of the input and output whereas that of quantity measures how much output (gain) with how little input (cost).
3. Benefit is regarded as the integration of efficiency, utility, and effect.

According to the above interpretations, interest can be divided into two forms of the absolute and the relative benefit, namely

$$\text{Financial benefit} = \text{output (gain)} - \text{input (cost)}$$

or

$$\text{Financial benefit} = \text{output (gain)}/\text{input (cost)}$$

4.2.2 Technology Transfer Benefit

Chinese scholar Chuanqi He divided the benefit of a system into four parts, namely direct, indirect, economic, and social benefit, which is illustrated in Figure 4.1.

The benefit of economic activities can be classified as direct and indirect benefits based on the direct economic results in activities of the system. Direct benefit refers to the financial benefit which is the result of economic activities and obtained immediately after these activities, whereas indirect benefit is the financial benefit which is gained under the impact of economic activities. Technology is bound to have a huge impact on the regions where technology is transferred and on the process in industries in the region. Consequently, financial benefit can be classified

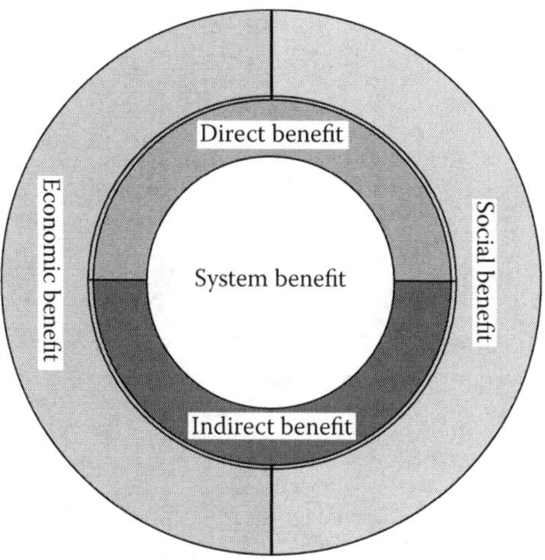

Figure 4.1 System benefits.

as practical and social financial benefit which reflects the improvements in personnel techniques and employment.

In the market mechanism, technology transfer is driven by economics. Technology transfer, in some sense, can be regarded as a kind of reaction for better interest opportunities and economic space when part of the technology functions in the market mechanism. As technology transfer is related closely to economy, enterprises or industries place emphasis on improvements in competence and efficiency, which results in unexpected indirect benefits, especially social benefits. It is local societies that gain advantage from these two kinds of interest whose cost is naturally paid by governments.

4.2.3 Relation between Technology Transfer Cost and Its Benefit

Technology transfer cost is discussed here to compare it with its benefit, that is, the expense incurred for technology transfer is for gaining benefit from it.

Technology transfer cost and its benefit, being inseparable, are two basic aspects in the study of technology transfer economy. There are several relationships that follow:

1. Mutual causality between technology transfer cost and benefit
   Technology transfer involves two sides and can be regarded as cost to one side and benefit to another. In addition, when it comes to the transfer of oligarchic and nonoligarchic technology, the cost of oligarchic technology transfer can convert into nonoligarchic interest; and vice versa.

2. The contradiction between technology transfer cost and benefit
   Technology transfer benefit is a function of its cost, with cost being the independent and interest the dependent variable, wherein lies the contradiction. Although the benefit may increase with the increase of input in some situations, as cost rises continuously, the benefits decrease and the efficiency of cost utilization drops or even negative correlation occurs. Therefore, the relation between the two should be studied in greater depth to optimize the utilization of cost before technology transfer.
3. The connection between technology transfer cost and benefit
   Technology transfer cost and its benefit cannot be studied in isolation. Transfer cost alone can only reflect the value amount for transfer, for example, the value of technology A is RMB 300K and that of technology B is RMB 200K. It is hard to decide, on the basis of only the technology transfer cost, whether RMB 100K more should be spent on technology A rather than B or whether it is reasonable to expect that technology B can save RMB 100K more than technology A. Therefore, technology transfer benefit should also be taken into account. Similarly, it is impossible to study technology transfer benefit without its cost because on one hand, the value of transfer cost, being a dependent variable, cannot be calculated without its cost which is an independent variable; on the other hand, it will be of little use and a waste of limited resources if the benefit is marginal and if the increase of such benefit is based on its cost on a large scale.
4. Fuzziness of technology transfer cost and benefit
   The fuzziness of transfer benefit can be inferred from an evaluation of the achievement value resulting from the spillover of technology transfer, such as improvement in employment, in the level of labor skills, and in the upgrade of knowledge level, all of which can hardly be valued in money. At the same time, this fuzziness can be reckoned by experts in different departments from their perspectives. It can also be reflected in many more aspects which include not only the uncertainty of technology transfer, such as the level of absorption and transition in the process, but also uncertainties of resource consumption and sense of identity by different people. Consequently, the fuzziness can be seen from the perspectives of both technology transfer cost and its benefit.

4.3 Measurement of Technology Transfer Cost and Benefit

The technology transfer cost and benefit vary in their technical factors but agree in the way of measurement. In this chapter, transfer cost and benefit measurements are illustrated with the example of technology talent and production techniques (in the sense of craft).

4.3.1 Measurement of Technology Talent Transfer

The cost of technology talent transfer involves all the expenses in the process from the occurrence of talent transfer (voluntary or involuntary) to the substitution of the talent by managers; this can be studied in two aspects.

With regard to technology talent, the transfer cost involves transfer expenses (contract breaking fees, commission charge, part of re-employment training expenses paid by trainees, income loss due to the constraint of time during the transfer, migrating fees due to transfer to a different place), loss of formal social capital (the relationship established in the formal workplace), family pressures the lack of support from families and relatives), and psychological pressure the loss of being identified by organizations).

As far as enterprises are concerned, the transfer cost of technology talent includes tangible and intangible costs. According to the survey made by the American Management Associations (AMA), the cost of one substitution for an employee, which involves the cost from the beginning of the talent transfer to the end of substitution, makes up at least 30 percent of his or her annual salary and 1.5 times or higher in positions that are in short supply of skilled employees. This phenomenon can be explained not only by tangible cost, which occurs directly during talent substitutions including the expenses of advertising, interview, and training, but also by the intangible cost, which has little direct link with fees actually paid and generally includes management expenses and the decrease in labor efficiency. The talent transfer cost is illustrated in Figure 4.2.

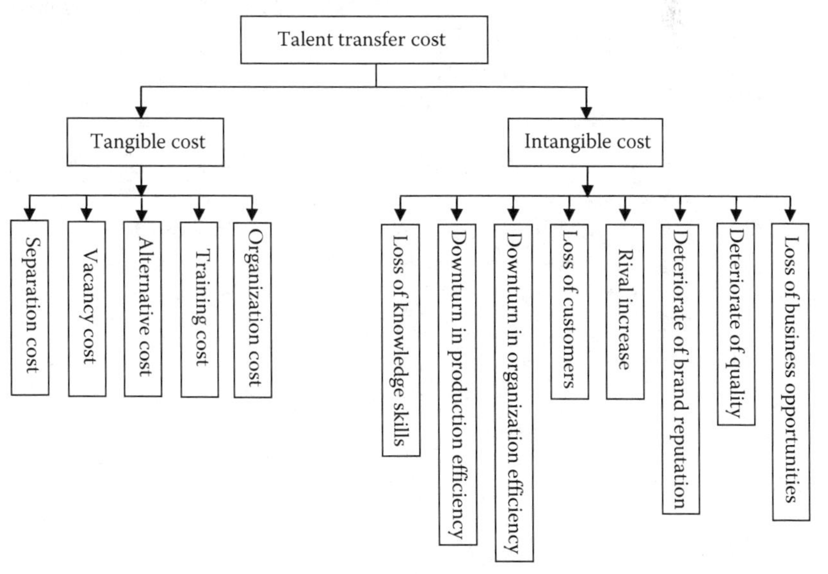

Figure 4.2 Cost analysis on talent transfer.

4.3.1.1 Cost Evaluation Model of Technology Talents

Whatever the reason, technology talent transfer occurs when there is a certain difference between self expectation and actual gain or when their talent cannot be exploited completely.

The value of self expectation is assumed to be V_0, which mainly reflects the demand for salaries and privileges. In modern societies, however, influenced by many factors such as constraints from outer environment and improper decision-making, the value of self expectation cannot be realized completely. The value realized can be assumed to be V_1; in this case, the difference between V_0 and V_1 can represent the psychological drop, which is denoted as ΔV (shown in Equation 4.1)

$$\Delta V = V_0 - V_1 \tag{4.1}$$

Obviously, it is possible for talent to be transferred when $\Delta V > 0$, which is the premise of this transfer. ΔV is highly related to the anticipations of talent. Even when the talent is well realized, namely V_1 is large enough, such transfer still occurs as these talents have high expectations so that $\Delta V > 0$ still exists.

On some occasions, such transfer may not occur even when $\Delta V > 0$ exists. According to the economic-person hypothesis, these talents should consider not only possible interest V_0 (assuming all the values can be realized after the transfer) but also the cost. The transfer only occurs when the interest surpasses its cost.

Assuming the expected interest after transfer is V' and the possibility of success in the transfer is P, according to the formula of expectancy theory (Victor H. Vroom), M (motivation) = V (valence) $\times$ E (expectancy), the formula of talent transfer motivation can be expressed as in Equation 4.2.

$$M = F(V') \times P \tag{4.2}$$

The motivation of talent transfer results from the subjective decision-making in the form of the product of V' and P. The transfers are unlikely to be made if these talents are not confident of their efforts (namely the possibility of transfer is little) even if the incentives are great. The desire for transfer can be strong only when V' and P are considerably high, which will assure that M is high enough.

C denotes the talent transfer cost, as mentioned above, including the transfer expenses, formal loss of social capital, family pressure, and psychological pressure. The relative cost-benefit ratio of talent transfer can be expressed by Equation 4.3 if motivation M and cost C are given,

$$R = M/C \quad (R > 0) \tag{4.3}$$

The talent transfer occurs only when $R > 1$ and $\Delta V > 0$ simultaneously. That is to say, when the talents are improperly used the will to transfer arises and the benefit of such transfer surpasses its cost, only then can the talent transfer take place in essence.

After roughly estimating the range of transfer cost according to whether talents choose to transfer, it is clear that $C < M$, which is shown in Equation 4.4.

$$C < F(V') * P \tag{4.4}$$

4.3.1.2 Enterprise-Based Cost Measurement Model

As mentioned above, there are not only tangible but also intangible costs during transfer of technology between companies. The methods of calculation vary due to their different natures. In this case, the separating aggregation method is applied here, which means measuring tangible costs with direct calculation first, then evaluating the intangible cost by fuzzy evaluation method, and aggregating the costs.

1. Calculation of tangible cost by the additive method (T)
   Organization cost (O') This refers to the loss of value during examination and employment, selection, transfer to new posts, including position analysis, compilation integration, personnel demand forecast, and operation costs of register notice, qualification check, exams, physical examination, assessment, public summons, and appointment.
   - Separation cost (S'). It includes negotiation-transferring cost (expenses of face-to-face conversations with talents that are to be transferred), managerial cost in relation to transfer (managerial expenses in dealing with talent files, salary, and welfare), and transfer cost (subsidies paid to talents who are to be transferred).
   - Vacancy cost (V'). It refers to expenses related to a temporary position vacancy caused by the transfer, such as overtime pay, expenses paid as salaries to temporary employees, outsourcing fees, and expenses for supervision and work rearrangement.
   - Replacement cost (R'). It includes the fees for processing related information, time cost of supervising staff, and performance differences between new and old employees.
   - Training cost (Tr'). It includes expenses in adjusting to the new position, skills training cost of new technology talents (such as on-the-spot training, training before employment, and vacancy studies).

   The tangible cost of technology talent transfer in companies can be computed if all the five kinds of cost are put together, as shown in Equation 4.5.

$$T = O' + S' + V' + R' + Tr' \tag{4.5}$$

2. Calculation of intangible cost by fuzzy evaluation
Because the fuzziness can be reflected in the loss of knowledge skills, downturn in productivity, and organization efficiency as well as loss of customers, the intangible cost can be valued by means of fuzzy evaluation. The procedural details are shown as follows:
    - The establishment of a panel. The panel can be formed by experts from outside and internal staff from different hierarchies.
    - Determination of fuzzy evaluation factor set $U \cdot U =$ [the loss of knowledge skills, downturn in productivity and organization efficiency, loss of customers, increase in the number of rivals, deterioration in brand reputation, deterioration in quality, loss of business opportunities].
    - Determination of evaluation level set $V \cdot V =$ [very important, important, slightly important, ordinary, and not important].
    - Evaluation of elements in the set U according to the comments in the set V made by the panels.

    Fuzzy matrixes in reflection of the fuzzy relation between U and V aid in the calculation.
    - The weight allocation set A is determined by rating the importance of elements in U after discussion by panels. It should be emphasized that the total of all the weights is 1.
    - Fuzzy evaluation models and comprehensive evaluation is established as B on all the elements.
    - Fuzzy transformation is $B = A \times R \cdot B$ denotes comprehensive evaluation results on the elements of set V.
    - The comprehensive evaluation levels of factors in the intangible cost can be made with these steps above. Thereafter, the ratio of the intangible part in the total cost can be made by quantifying these levels with the panels. Finally, the estimated value of the intangible cost (I) can be made.
3. Summarizing cost of technology talent transfer in business
The total cost of technology talent transfer can be made by putting the tangible and intangible cost together as in Equation 4.6.

$$TC = T + I \tag{4.6}$$

4.3.1.3 Example Study

Assuming a technology talent, considered as technical talent, is transferred from company A to B, the transfer cost indices are shown in Table 4.1 (the unit is RMB in thousand). In this case, the cost of technology talent transfer can be estimated from the point of view of the enterprise.

Table 4.1 Tangible Cost Indices in Technology Talent Transfer

O'	S'	V'	R'	Tr'
1	3	30	100	4

According to Equation 4.5, the tangible cost of technology talent transfer is expressed as follows:

$$T = O' + S' + V' + R' + Tr' = 138$$

Then, the intangible cost of technology talent transfer can be calculated with the fuzzy evaluation method provided the sets U and V mentioned above are given.

The fuzzy matrix R can be created after evaluation of all the elements in U with reference to comment set V by ten members of the staff, in the company, who are familiar with the talent.

$$R = \begin{bmatrix} 0.4 & 0.3 & 0.2 & 0.1 & 0.0 \\ 0.1 & 0.4 & 0.4 & 0.1 & 0.0 \\ 0.0 & 0.0 & 0.1 & 0.4 & 0.5 \\ 0.0 & 0.1 & 0.2 & 0.5 & 0.2 \\ 0.1 & 0.1 & 0.6 & 0.2 & 0.0 \\ 0.0 & 0.0 & 0.0 & 0.3 & 0.7 \\ 0.1 & 0.3 & 0.3 & 0.2 & 0.1 \\ 0.0 & 0.1 & 0.1 & 0.2 & 0.6 \end{bmatrix}$$

Then, the fuzzy transformation $B = A \times R$ can be made by confirming the weight allocation set $A = [0.15 \quad 0.05 \quad 0.05 \quad 0.15 \quad 0.05 \quad 0.25 \quad 0.20 \quad 0.10]$

$$B = [0.090 \quad 0.155 \quad 0.185 \quad 0.260 \quad 0.310]$$

$$I = 100 \times B \times V = 100 \times [0.090 \quad 0.155 \quad 0.185 \quad 0.260 \quad 0.310] \times \begin{bmatrix} 9 \\ 7 \\ 5 \\ 3 \\ 1 \end{bmatrix} = 391$$

After discussion with the panel, the ratio of the intangible part to the total cost can be made by proper quantification of evaluation levels. The intangible cost I can then be obtained.

Therefore, the total cost here is $TC = T + I = 138 + 391 = 529$ (thousand RMB)

4.3.2 Measurement of the Cost of Production Technology Transfer

The transfer cost of production technology varies in relation to the two parties of production technology.

4.3.2.1 Factors That Affect the Transfer Cost

1. The benefit obtained by the production technology
   The most important benefits from production technology transfer are the economic and the social benefits, which are brought about by the transfer; the transfer cost must depend on use-value of the technology. Furthermore, production technology transfer cost depends mainly on the anticipated benefit when it has been transferred, comprising the higher labor productivity, improvement in product quality, saving of resource consumption, and cost saving level, all of which result from the technology transfer; the life cycle of product reform and market share after the development of the new product; improvement in business management, product market, managerial environment, the increase of enterprise vitality, safeguards, etc. The quantitative values of direct and indirect benefits can be deduced by these predictions.
2. Cost of research and development
   The cost of research and development of the new technology by transferors includes direct and indirect benefits. Direct benefit comprises material cost, specialized equipment expenditure, labor cost, peripheral cooperation fees, training and advisory fees, labor protection necessities, transport and storage, technical archive management, bank credit and so on, whereas indirect benefit comprises the management fees, fixed assets in the process of technology development (e.g., depreciation expense of equipments, factories, and instruments and related public facility expenditure).
3. Corollary investment after production technology transfer
   Because the technology products are high in specificity, the equipments of transferees, their manufacturing process and production scale should match the technology transferred to be well used. These investments include modification of old equipments, expenses incurred in purchasing corollary facilities, purchasing production sites, and construction of new factories, and employing and training fees. In case the trend of profit increase remains unchanged, more corollary investments are made leading to increase in total investment, and higher risks, which in turn affects transfer cost of production technology.
4. The license of production technology transfer
   The license of production technology transfer can be divided into five kinds. (1) Exclusive license, which means transferees share the right of obtaining techniques in the contracts exclusively in the fixed period and area where

transferors and third parties have no right to produce and sell goods with these techniques. (2) Sole license (mono-license), which means transferees, as well as transferors, rather than third parties, share the right to manufacture or sell using techniques permitted in the contract during the fixed period and in some specific area. (3) General license, which means transferees as well as transferors can not only have the right of using the techniques mentioned in the contract but also transfer them to third parties during the fixed period and in some specific area. (4) Affiliated license, which can also be classified as sub license or transferable license. Affiliated license means that transferees can earn money from the techniques transferred by signing contracts independently with third parties in some fixed period and in some specific area. (5) Cross license, which means both the sides, transferors and transferees, can get access to the techniques of their counterparts by signing contracts. These five kinds of transfers are accompanied by their respective kinds of cost.

5. Expenses of production technology transfer
   Transfer fees refer to all the expenses paid by transferees in signing and executing the contracts, which include fees for feasibility study such as installation, debugging, training, and market exploration; travel and managerial cost during negotiations such as fees for accommodation and commuting; some legislation expenses such as fees for treaty consulting, review, and registration; other expenses concerned with executing contracts such as entertainment expenses for inviting transferors and commission charge for agents. These expenses vary in size depending on the complexity of projects and period of negotiations as well as the contracts.
6. Advancement and maturity
   The advancement of production technology is an important criterion for transfer, which can be placed at different levels—of having originated in the outside world, being a leading technology worldwide, of having originated domestically, and being a leading technology domestically. The more advanced the technology products are, the larger the quantity of condensed social labor will be, and the more it will be difficult to research and imitate the technology resulting in fewer rivals and higher prices.

   The benefit and risk of investment varies with the degree of maturity in the technology, which means the factors of prices in production technology varies with the degree of development and technical maturity. In general, the maturity of technology products may result in lower risk for the transferees, shorter investment periods, less capital burden, and increase in their cost.
7. Life span of production technology and stages of technology in the span
   As the life span of technology differs, the cost allocated annually is high for techniques with a short life span, and vice versa. The technology may go through three stages in its life span, namely the growing, the mature, and aging stages. The techniques decrease with time.

8. Supply and demand in technology transfer market
   This factor can be reviewed from two sides—one being the supply and demand relationship in the technology market, the other being the relationship in the materialized commodity market. As for the transferors, the materialized products are the deciding factors to buy technology products and to determine the prices. The law of value works by the supply–demand relationship in the materialized product market.
9. The amount of information acquired by both sides of the transfer
   In general, transferees have little knowledge of the technology and the market and are uncertain about the economic benefit that will result from the transfer. The bidding prices may be reasonable if the transferees have the necessary knowledge and related information and are efficient in the absorption of the technology. On the contrary, the bids may be blind if the transferees are uncertain about the economic benefit that will result from the transfer. Only complete awareness of the standard and the performance of the technology products can help them make a reasonable bid.
10. The risk of production technology
   The risks of production technology arise from the comprehensive effect of all the price factors as well as others. These factors include amount invested in the production technology transfer, the investment recovery period, the technological maturity, management and operation after the transfer, the understanding of the transferees of the technology, information on the redevelopment, the start, operation, and share of market after transferring to productivity, and the number of competitors. The operative conditions, the technical and economic life span of the production technology, and the mode of payment for the price of technology are also taken into account.
11. Industry codes, international practice, and other factors
   There are some codes and practices on the transfer fees in different departments and products worldwide, shown mostly as the royalty rate. For example, the royalty rate in pharmaceutical technology is 10 percent to 15 percent, that on necessities being more or less 2 percent. These codes and practices may have a considerable impact on the cost of production technology transfer

4.3.2.2 Transfer Cost Estimation Model for Transferors

According to the analysis above, the transfer cost for transferors is shown by Equation 4.7.

$$C_{\mathrm{fo}} = \frac{C_{\mathrm{d}} + C_{\mathrm{o1}} + P_{\mathrm{c1}}}{n(1 - p_1)} + S\alpha\beta Y_m l \qquad (4.7)$$

where

C_{fo} ($C_{fo} = C_{flowout}$) is the transfer cost for the transferors

C_d ($C_d = C_{develpment\ and\ research}$) is the technology research cost, if the research period extended for more than a year, the cost should be translated into cost during the transfer, as $C_d = C_0(1+i)^k + C_1(1+i)^{k-1} + \cdots + C_k$, ($C_0, C_1, \ldots, C_k$) being the cost before the transfer

C_{o1} ($C_{o1} = C_{opportunity1}$) is the opportunity cost of development for this technology

P_{c1} ($P_{c1} = P_{currency1}$) is the cost for the transferors for transferring the technology, such as peripheral cooperation fee, advisory fees, transport and storage fees as well as the cost of using public facilities

n is the excepted transfer times

p_1 is the development risk rate of this technology

S is the production technology market capacity coefficient, $0 \leq S \leq 1$

α is the monopoly coefficient

β is the production technology contribution rate for acquiring profits

Y_m is the sum of increased benefit estimated after transfer

l is the equivalent coefficient of production technology outcome period, l = ratio of commercial period of prediction/legal patent protection period of outcome

4.3.2.3 *Measurement Model of the Transfer Cost for the Transferees*

According to the analysis of transfer cost for the transferees, the transaction prices of the transfer should be studied first. The transferors take into account the market situation of supply and demand for the two sides at this time, as well as the stage in the production technology's life cycle, to determine a profit-sharing coefficient r, and the necessary tax rate t, and the transaction price for the two sides can be shown as in Equation 4.8.

$$P_b = (1+r) \cdot C_{fi} \cdot (1+t) \tag{4.8}$$

It can be seen that the transfer cost of transferees is given by Equation 4.9.

$$C_{fi} = \frac{P_b \cdot (1+\delta) + C_{o2} + P_{c2}}{1-p_2} \tag{4.9}$$

That is

$$C_{fi} = \frac{(1+r) \cdot C_{fo} \cdot (1+t)(1+\delta) + C_{o2} + P_{c2}}{1-p_2} \tag{4.10}$$

where

C_{fi} ($C_{fi} = C_{\text{flow into}}$), is the transfer cost of transferees

δ is the disparity factor of technology between the transferor and the transferee, when the transferred technology is not the same kind as the transferor's leading technology

C_{o2} ($C_{o2} = C_{\text{opportunity2}}$) is the opportunity cost of transfer in this technique

P_{c2} ($P_{c2} = P_{\text{currency2}}$) is the cost of transferring technology, such as peripheral cooperation fees, advisory fees, transport and storage fees as well as the cost of using the public facilities

p_2 is the risk rate of developing this technology

4.3.2.4 Case Study

Assuming that a production technique flows from one part to another, the specific number is shown in Tables 4.2 and 4.3 (funds unit: RMB in millions), the transfer cost between the inflow part and outflow part in these conditions can be calculated.

It can be seen from Equations 4.1 and 4.2 that the cost paid by transferors and transferees for technology transfer are respectively

$$C_{fo} = (C_d + C_{o1} + P_{c1})/[n(1 - p_1)] + S\alpha\beta Y_m l = 238.75\text{ths.}$$

$$C_{fi} = [(1 + r)\cdot C_{fo}\cdot(1 + t)(1 + \delta) + C_{o2} + P_{c2}]/(1 - p_2) = 375.58\text{ths.}$$

4.3.3 Measurement of Benefit in Technology Talent Transfer

4.3.3.1 Benefit of Technology Talent Transfer

In a narrow sense, the benefit of technology talent transfer may involve "transfer benefit of talent (in groups) resources" and "benefit of talents (in individuals)." The former one is used in macro studies while the latter is used in micro studies.

Table 4.2 The Transferor's Cost Consumption Indicators

C_d	C_{o1}	P_{c1}	n	p_1	S	α	β	Y_m	l
10	2	1.6	1	0.15	0.7	0.3	0.5	30	1.5

Table 4.3 The Transferee's Cost Consumption Indicators

r	t	δ	C_{o2}	P_{c2}	p_2
0.15	0	0.2	1.5	2.1	0.1

The technology talent transfer refers to not only the transfer of talents, but also, in the sense of knowledge flow, the transfer of knowledge, technology, and creative ideas. In view of the economic growth, the transfer of talents obviously promotes the development of market economy and economic growth.

The greatest function of talent transfer lies in the way of knowledge transfer. It is the talents that transfer their knowledge and release their techniques, resulting in the improvement of knowledge in social organizations, the creativity during the transfer, and the continuous financial benefit by the application of techniques. Such benefit in talent transfer is brought about by the creation and appreciation of knowledge flow in essence accompanying the transfer, which can in essence be the actual meaning and value of talent transfer.

In this case, the benefit of technology talent transfer is defined as new economic and social benefits caused by its appreciation in the case of knowledge innovations by talents and technology applications in the process of orderly transfer in talent resources.

The main benefits of technology talent transfer include

1. Direct benefit which is the sum of the differences between the formal salary and new income of each of the employees' in their new occupations.
2. Indirect benefit which refers partly to the savings in employees' expenses owing to the conveniences in the new working environment, such as advanced utilities.
3. Psychological benefit which includes noneconomic utilities due to the alteration of positions, such as an ideal working environment, rise in social status, and the reunion of families.

4.3.3.2 Measurement Model of the Benefit in Technology Talent Transfer

By analysis of the Cobb–Douglas function, the economic growth factors due to the increase in talent capital can be separated to study the change in the economic increment caused by talent transfer.

In this case, the Cobb–Douglas function of economic growth in a certain region is shown in Equation 4.11.

$$Y = AK^{\alpha}L^{\beta} \quad (\alpha + \beta = 1) \tag{4.11}$$

where

Y denotes output
K denotes capital input
L is labor input
A denotes technology status
α is output elasticity of capital
β refers to output elasticity of labor

This is a linear homogeneous function, which is constant in returns to scale. This equation shows that the output is determined by three main factors, namely capital input, labor input, and technology status.

By separating increments in the Cobb–Douglas function, the financial benefit of talent transfer (FY) can be calculated as follows:

$$\mathrm{FY} = \beta_i \frac{\Delta L_i}{L_{i-1}} \bigg/ \frac{\Delta Y_i}{Y_{i-1}} \times \Delta Y_i \tag{4.12}$$

The premise of this model is

1. Technology talents transferred from the same region are homogeneous.
2. The capital of technology talents vary from region to region.
3. The talents transferred are made good use of.
4. The user benefit of talent transfer can be shown one year after the transfer due to the time-delay of talents' effect.

4.3.4 Measurement of Benefit in Production Technology Transfer

Compared with other transfers of production factors, the production technology transfer plays an even more important role in the economic growth of one nation or region. In developed countries, the economic growth depends on breakthroughs in technology, which contribute 70 percent of the growth, and the efficient applications of technical achievements. Even in developing countries, the economic growth increasingly relies on breakthroughs in technology and efficient applications.

4.3.4.1 Benefit of Production Technology Transfer

The benefit of production technology transfer refers to the effects on instruction, promotion, development, and growth of domestic economies, which indicates the impact of technology transfer on economic value.

The establishment of the process of technology flow and improving the technology transfer efficiency are rather critical issues and important reference points for technology transfer policy-making.

In the early years of international technology transfer, many countries made too many unilateral decisions to protect their initiatives during the transfer in the interests of their own economies. This resulted in the blocking of technology flow. However, the transfer process in recent times has been well regulated by mature international legislations even while the freedom of transfer flow is well guaranteed.

The technology carriers can be protected in different forms such as patents, copyrights, transfer and warrant of license, transfer of human resources, and merging or separation of different technology flows, which have a direct link to the final results of technology transfer. The application of technology flow can affect the efficiency of transfer significantly.

It should be noticed that a large amount of technology transfer is achieved by means of informal technology flow, such as publication of archives, transfer of technical employees, etc. On one hand, these means of transfer are low in cost and well absorbed, resulting in high transfer efficiency; on the other hand, transferees can not only receive all the techniques passively, but also absorb some of them selectively and develop on those chosen thereby improving the transfer benefit.

4.3.4.2 Measurement of Production Technology Transfer

Production technology and technology flow are closely related. Technology relies on the transfer, introduction, transmission, and trade among entities at different ranges and levels, constituting a complex "technology transfer network." Technology transfer entities, techniques, and technology flows are three components of production technology transfer, as shown in Figure 4.3.

As is illustrated in this network, the technology transfer is achieved by a series of procedures each of which has a different kind of impact on the economy, resulting in different criteria for each of these procedures. Although the impact does exist, it is by no means an easy task to find a clear relationship between these procedures and the economy. Some useful information from economic phenomena can be obtained when technology transfer has been achieved over a long period of time (several months or several years) but the phenomena studied may be affected by other

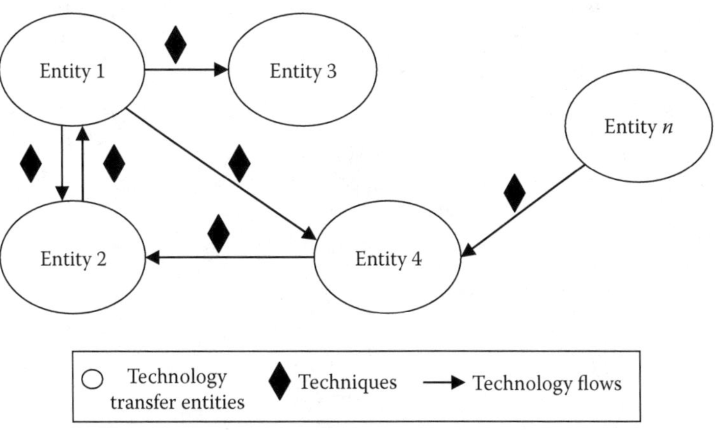

Figure 4.3 Network of production technology transfer.

factors. In conclusion, the precise measurement of financial benefit in production technology transfer can hardly be realized.

To study the impact of technology transfer on the economy, three factors in the network of production technology transfer, namely technology transfer entities, techniques, and technology flows are evaluated here.

Assuming a technical potential η and economic value per potential λ, then the financial benefit π of this technique can be expressed as follows:

$$\pi = \lambda \times \eta \tag{4.13}$$

After the research and development following the transfer, the new technical potential is η, while the change rates in technical potential, economic value per technical potential and technology economic value are as follows:

$$\eta = \Delta\eta/\eta, \quad \lambda = \Delta\lambda/\lambda, \quad \pi = \Delta\pi \times \pi \tag{4.14}$$

Obviously,

$$\pi + \Delta\pi = (\lambda + \Delta\lambda) \times (\eta + \Delta\eta) = \lambda \times \eta + \Delta\lambda \times \eta + \Delta\eta \times \lambda + \Delta\lambda \times \Delta\eta.$$

In this case, the benefit of production technology transfer is known.

$$\Delta\pi = \Delta\lambda \times \eta + \Delta\eta \times \lambda + \Delta\lambda \times \Delta\eta.$$

Chapter 5

Technology Transfer Analysis Based on Technical Diffusion Field Theory

Any innovation in science and technology will diffuse once it is publicized. Since World War II there has been an international trend to diffuse technology; spontaneous at first, with the United States and Germany spreading the relatively low-tech and labor-intensified technologies into Japan and the Four Little Dragons, namely Singapore, Hong Kong, Taiwan, and South Korea. The same thing happened in the 1970s, but from some developed countries to the developing countries on an international scale with specific purpose. This increasing diffusion of technology exerts great influence on industry structure, economic policy and development, social, and cultural life.

5.1 Technical Diffusion Field Statement

5.1.1 Technical Diffusion Field

The term "field" refers to the distribution status of a physical quantity in space. In natural science, "field" is commonly used to study laws of space distribution and changes in some physical quantities such as temperature, density, potential, power, and speed. In the study of technical diffusion also, "field" can be used to describe the distribution of and changes in technical diffusion.

In essence, technical diffusion is the process of transforming technology into social production and reproduction directly. This is not only among countries and regions but also among different industries as a result of unbalanced technology development. Thus, it is a multilevel and multi-dimensional diffusion. Like in many science and technology problems, it is necessary to be concerned about the distribution in space and changes in some physical quantities (such as temperature, density, potential, power, speed, etc.); the concept of "field" and the idea of technical diffusion field to reveal these laws and provide a theory and method for the study on mechanism of technical diffusion and absorption is therefore introduced.

Technical diffusion field is a special form of material and information based on technical diffusion, through which the technology bearers interact.

Under certain conditions, technology bearers act as the field source, which stimulates the technical diffusion field to spread the technologies constantly. When the source has a higher level of technology than the receiver in the diffusion field, it will have an effect on the latter, though this effect may be zero. The original location of this technology is known as the technical diffusion source and the receiving technologies as the technical absorbers. The direction of influence that the technical diffusion source has on the absorber is called the technical diffusion operational direction, which points from the source to the absorber. As the kind of effect varies among different technical absorbers, technical diffusion has different sizes.

5.1.2 Technical Diffusion Process

A product system should integrate the knowledge of reification, literature, and connotation. Technical diffusion includes the following two aspects: software and hardware diffusion, which means the diffusion of literature and connotation and the diffusion of reification correspondingly.

Different forms of technology can diffuse in two ways: the nonproperty form and the property form. The first can be purchased easily without any special restrictions while the latter, like secrets of enterprises, has limiting conditions.

Technology diffusion is a complex process which can be divided into three intercross phases.

5.1.2.1 Plan and Choice

The technology supplier directly or indirectly contacts the enterprises with technology requirements through various information channels setting in motion the process of accepting the idea of technical diffusion. This is the "software" diffusion phase, which is the diffusion of literature and connotation.

The plan and choice phase starts from confirmation of the need for some technology, often decided by enterprises, but the initiative may derive from other sources; foreign enterprises with overseas investment, for instance. During this phase, it will decide which technology to import, through which kind of mechanism (import equipments, license or foreign investment, etc.), and the resource of the technology.

5.1.2.2 Digestion and Absorption

Digestion and absorption are the processes by which the absorber acquires innovative technology legally when the information is transferred from the technical diffusion source. This acquisition varies in form, independence, possession, and utilization.

Digestion and absorption constitute the "software" diffusion phase where skills and hardware are diffused by literature and connotation and the technical diffusion of reification is capital in nature.

We can take this phase in the process of technical diffusion as the conception process which actually equals the innovative technical diffusion.

W. Haig, Senior Researcher at the Stanford International Consultation Research Institute, argued that for a country or an enterprise, technology transfer can create more knowledge. However, it cannot promote the economic development directly. It is necessary to undertake other activities to create an efficient environment and conditions to accept these technologies. On the other hand, it has to inspire a multitude of departments, which may help to attain the result with half the effort so that the technology can help companies or the country to promote saving in power, cost and resources, and improve economic growth. If this technology remains static after being introduced, it is not an effective technology import. The critical thing, when introducing technology, is to keep the innovative spirit alive. Only with the innovative spirit, can technology introduction be meaningful and constantly developing. The direct effect of technology introduction is to create the technical potential, but other conditions are needed to generate economic benefit. Digestion and absorption are the processes by which the value of the introduced technology is increased by the essence of innovation.

The Japanese expert of technology transfer theory, Yu Saito wrote in "Technology Transfer Theory: Improve all the introduced technologies frequently to adapt to the system in Japan, which is the characteristic of Japanese technology transfer" and "the introduced technology is not necessarily to be used by relevant enterprises and stays unchanged. Thus, it should attach great attention to get immobilized when transferring technologies."

Japanese scholar Wu Lin argued, in an article on Nippon's experiences in introducing and digesting advanced technology in the Third World, that the transformation of introduced technology into an independent development process can be divided into five phases, namely, operation, maintenance, repair (small improvement), design, self-production (technology system management). Hoshino divided the above five phases into three steps: imitation, partial improvement, and comprehensive improvement.

The digestion and absorption should be considered as follows. It is a process of studying and digesting the introduced technology, and integrating it with the local economy. The process of digestion is to analyze, understand, and master the introduced technology while the absorption process is to improve, innovate, and promote the digested technology, and then to apply it in actual production for achieving productivity. Digestion and absorption can be divided into four stages.

1. *Use*: This stage involves applying introduced technologies correctly and giving full play to them to meet the original design, and being able to carry out the necessary simple repairs at the same time.
2. *Grasp*: This stage ensures an anatomical understanding of the new product technology or technique that is a decomposition and combination of the technology system.
3. *Innovation*: For technology absorbers, the introduction of technology is merely to aid innovation rather than a ready-to-use mature technology. Introduced technologies will usually generate three stages when developed further by innovation–first, a partial reform of the original technology to adapt it to the new technological environment and improvement to its theoretical structure and formation; second, transplantation, that is applying the original or the digested technology to other situations and expanding its scope of application; third, integration, that is integrating introduced technologies and domestic advanced technology into a new technology system. Innovation embodies the essence of digestion and absorption.
4. *Promotion*: Innovation is not the end of digestion and absorption but merely the beginning to promote greater technological achievements.

As the technology is always developed for specific conditions, it is often necessary to make some adjustments to it different from its initial design when transferring it to a new environment so that the technology can be successfully digested and absorbed. This is true for almost all the introduced technologies before they melt into the local economy. When the introduced project is complex, adaptive adjustments start at the very beginning from the design of the factory and the equipments, to the whole process of construction, installation, and troubleshooting. The operative effects of the introduced production system, such as the rate of output, product quality, product unit costs, will improve as business experience accumulates. In addition, the design and operation improvements brought about by the adjustments and absorption process, for example, a good program which can save time by improving the order of operation in quality control and by the use of abandoned components, can introduce a production system which is better than the original in efficiency and quality of output.

The stage of digestion and absorption ends when the introduced technology is digested and integrated with the local economy successfully. It can be concluded that this stage has been achieved when actual output meets or exceeds the original design level, and production costs are market competitive, and production systems integrate with the local economy.

5.1.2.3 Diffusion

Technical diffusion is to diffuse the introduced technology or technical knowledge, separate from the material form, from enterprises that use the introduced technology to other sectors, for example to other enterprises in this industry (domestic or foreign), to raw materials suppliers and to other sectors of the national economy.

Diffusion takes place when technicians or skilled workers move from the company, where the technologies were introduced, to work in other companies. Technical diffusion can be seen in two scenarios.

1. Companies with technologies introduced from foreign countries try to help other companies (domestic and foreign) acquire them.
2. Companies, where technologies have been introduced, supply technique and technology services to raw material suppliers and collaborative companies to make raw materials, components, assets, and facilities locally.

The most important aspect of technical diffusion is its implementation, that is, the absorber adopts the necessary technologies, for example, engineering installing technology, product organization, and innovation technology. Technology diffusion involves the materialization of the technical absorber and the reification and absorption process in the technical absorber. The favorable implementation process of technical diffusion not only has a direct effect on the benefits to the technical absorber, but also has a direct effect on the digestion–absorption technology.

The process of technical diffusion is fraught with great uncertainty as even though the technical absorber has accepted the idea of technical diffusion several issues like the choice, extent, time, and methods of technology adoption still remain. However, during the process of technical diffusion, the absorber has a clear legal agreement. Both sides are bound by their agreements; on the whole, the implementation is certain. In the process, to master the construction and installation technology as well as production and manufacturing technology, it is necessary to train the personnel of the technical absorber and provide instructions from the technology diffusion source. The technical absorber should undertake adjustments, changes, or innovation of organization structure. The implementation of technical diffusion has a cumulative effect on the technical absorber. It is a gradual process on the basis of technology, organization, management, etc. The implementation of technical diffusion is dependent on time. When the anticipated technical economy effect is desirable at that point of time, the technical absorber will decide to adopt it immediately, or it will choose to wait.

Judging from the time sequence, the technical diffusion process can be segmented into four parts: adjustment and transformation of organizational structure, personnel training, debugging and installation of facilities, the use and study of the technology of production and manufacture.

5.1.3 Technical Diffusion Mechanism

5.1.3.1 Form and Implementation of Technical Diffusion Mechanism

Technical diffusion to the technical absorber happens in diverse ways. U.S. experts E.B. Skolnikov lists the following ways technical diffusion: permit (including the transfer of technology sales, complete sets of equipment, hardware sales, joint ventures, and the contract tender), patents (including publications, books, marketing brochures), interviews (including meetings, training, education, public policy discussions, debates), and information (including electronic, communications, intelligence). British researchers summarize the current methods as follows: transportation (like transportation of advanced industrial goods), migration (like expert migration), copy, transfer (like industry transfer), subcontract (production), development assistance, joint ventures, investment, patent and chartered (transfer), information exchange, personnel training and exchanges, and consultation and cooperation in research and development. An Indian scholar, Chaturvedi, summarizes the methods as follows: foreign investment, technology cooperation, provision of equipment, complete sets of contracted projects, government agreement, collective purchase, consultations (noncommercial), international academic activities, and personnel training and education.

5.1.3.2 Choice of Technical Diffusion Mechanism

The choice of transfer mechanisms depends on several factors like the motivation of the technology supplier, the technical feature, the local level of technical capabilities, the relative strengths of the two sides for negotiation, and technology recipients.

1. *Technology providers*: If a technology is not important to the technology provider, he is usually willing to supply the technology without any restrictions. However if it is very important to him because of its long-term competitiveness, especially technology that can serve as an effective barrier for other corporations, the provider will not be willing to transfer it, unless it can be confirmed that this transfer will not weaken his long-term competitiveness. When the transfer mechanism allows him to have a certain degree of control on the use of the technology, the provider will be willing to transfer relative technologies.

   One transfer mechanism, namely, individual proprietorship, is one way for providers to maintain control on the technology. A joint venture, with the provider holding the larger share of the capital indicates relative restrictions or limitations in the contract and permit contract, so as to make sure that the provider has control of the technology.

2. *The local level of technical capacity and technical features*: When the capacity of the technical recipient is close to that of the provider, independent technology permit is an effective mechanism for the transfer. When the technology is complicated, and technical level of the recipient is lower, technology permit may have lower benefit. To transfer technology successfully, the provider's positive cooperation is required. In this case it is better for the technical recipient to adopt the joint venture style as the transfer mechanism.
3. *The relative strength of the two negotiating parties*: When the technology has been in use for long and can be provided by many companies, the power of the supplier is weak and the transfer mechanism is likely to be one with no constraints in the permit process or one of outright sale. When the technology features high product specialization, is responsible for a great difference in the manufacture, and is the secret of or has been patented by a company, then the few businesses that provide this technology have a strong negotiation power. In this case the technology is likely to be adopted with a constrained permit, or foreign direct investment, or both, such as an integrated mechanism.

5.2 Analysis of the Influence Factors of Technology Transfer in the Technical Diffusion Field

5.2.1 Social Environment Factors

Every district has different needs for industrial products in developing the national economy. A developing industry is dependent on natural, economic, and technological conditions, labor resources, and people's culture, values, and historical development. The special conditions for different departments and industrial products also vary. This determines whether a technology is conducive to the development of the region, and whether the region should adopt it.

5.2.1.1 National Policy

The objective of the policies of technical diffusion for any enterprise is to ensure that it is the only way to maximize the benefits for the enterprise. In accordance with this objective, the policy should lay stress on scope integrity, incentives, and restrictions and follow the scientific nature of the elements in both the long-term and short-term policies. During the actual execution of the work, policies should focus on stability and coordination to give full play to the guiding principles and support.

The policies which influence technical diffusion include: organization and management policies, labor wage policy, technology commercialization, financial and taxation policy, financial and credit policy, and import and export trade policy, as well as the policy to promote scientific and patent information, the guiding principles of scientific research institutions and college schools, and the training policy for technical personnel and work staff. These specific policies create a sensitive technological atmosphere conducive to technical diffusion, help establish a scientific technical-evaluation system to improve the efficiency of the technology transfer and reform the education system to adapt to the needs of technical diffusion.

5.2.1.2 Nature of the Guide-into Region

1. *The impact of a market economy environment on technical diffusion*: In market economy, the market is the corporations' world. Market environment influences technical diffusion mainly through its system, patterns, mechanism, scale, and order. Market systems include the product market (material market of production and living) and factor market (financial markets, labor market, and technology market) primarily. Corporations reap excessive profits through technical diffusion. Market pattern refers to the seller's or buyer's market. Market mechanism refers to the price, competition, and risk mechanisms. Market scale refers to market demand quantity, supply quantity, volume maintenance and the scope of the market. Therefore, an open market economy environment is conducive to technical diffusion. However, it is not enough for economic regulation to use only the market lever; large-scale technical diffusion still needs the government's macroeconomic policies.
2. *The impact of digestion and absorption on technical diffusion*: Transfer diffusion must be in tune with independent research and development, should help the recipient digest and absorb the technologies and develop its own innovative technologies, enhance competitive motivation through technical diffusion to participate in the international market. Digestion and absorption need the union of enterprises and areas. This union and sharing of resources will generate the synergistic ability to form a strong digestion, absorption, and production capacity. For the development of education, the digestion and absorption capacities must be improved.
3. *Mind state of technical diffusion*: A good state of mind includes a thirst for knowledge, and healthy curiosity without xenophobia, which is a positive factor for quick adoption of new technology. Because of the imbalance in social, economic, and technological development and its own characteristic features, a region cannot always be in a position to innovate in all its unit technical groups. It is important to choose the correct region for technology transfer by proper research into its features.

5.2.1.3 Legal Environment

The scientific and technical legal system is an important component of China's legal system, including the system of scientific and technical law, implementation of scientific and technical law, research into scientific and technical law, and so on. Science and technology development and any enterprise's technological progress need legal protection with immediate effect. Establishing a perfect legal system and creating a good legal environment are powerful factors in promoting technological progress and diffusion.

5.2.1.4 Institutional Environment

Institutional environment includes the economic system, the science and technology system, polity system, and so on. Its role and impact on the technical diffusion of an enterprise strongly reflect the corporate structure and principal part of investment. Whether enterprises can become a principal part of technology is a fundamental problem which relates to the union between science and technology and production.

5.2.1.5 Location Environment and Local Industrial Structure

A good location environment is one that provides perfect infrastructure and a cooperative and convenient economy, which are necessary for an enterprise to achieve steady technological progress with these unique geographical advantages.

The correlation between industry structures impacts the efficiency of technical diffusion. The degree of technical contact and information exchange between departments are limiting factors that affect industrial growth. The improvement in technological innovation efficiency of one industry sector permeates into other sectors, which must be further diffused through conduction in industrial associations. Meanwhile, technical diffusion accelerates the progress of the adjustment between industrial structure and industrial innovation. The economic benefits of technical diffusion and industrial promotion can be achieved only by this.

5.2.1.6 Industrial Environment and Structure

The industry is the product of development in social labor and change in industrial structure. With improvement in industrial production and technical development, production socialization increases. Integrating all departments, enterprises and even all sectors of the same industry in the region as a whole will increase social benefits for the entire industry.

The industrial environment includes industry scale and status, competition and cooperation, development situation, and so on. The industrial environment's impact on the technological progress of an enterprise is reflected in the competition and collaboration. Strong product-match, collaboration, and fair competition among enterprises are necessary to improve enterprise industry environment. Therefore, it is imperative to establish mechanisms for equal competition, to strengthen industrial technology development and promotion, and to create a good industrial environment for technological progress.

The requirements, in terms of quality and quantity, for a certain technological industrial structure are different for different regions. In terms of quantity, the proportion of the relevant industrial structure of such technology is considered and in terms of quality, it is the internal environment in and adaptability of this region that is considered.

5.2.1.7 Financial Environment

Financial environment includes financial policy, financial system, financing channels, science and technology credit system, and so on. The financial environment is decided by policies and systems in force in the environment. In the present context of technological progress, technological invention, innovation, and diffusion require massive investments, and the returns are hysteretic and long-term in nature. Therefore, a good financial environment is particularly important for the development of enterprises.

5.2.1.8 Information Exchange System

An effective information exchange system can provide a correct and necessary basis for decision-making in technical diffusion. Technical diffusion is divided into two kinds depending on the diffusion taking place through intermediary services or otherwise. Intermediary services refer to organizational units and departments which act as agencies, bridges, or links between providers and demanders of new technological achievements. Based on the different service functions in technical diffusion, and the different relationships with innovation technology or technology diffusion sources, intermediaries can be divided into information intermediaries, innovation technology transfer agencies, and innovation incubators.

The characteristic of information intermediaries is that they generally do not have the innovative technology, but are in charge of the contact and communication of the technology according to demand and supply. Such institutions include nonprofit government agencies that promote scientific and technological achievements, free or paid services of the mass media, and civil specialized information technology advisory body. In the process of innovation technology diffusion, information intermediaries can help overcome the obstacles in the first stage and can provide basic information for those using innovation technology.

The innovation technology transfer agency serves transfer diffusion in special areas, and has a fixed or relatively stable relationship with the technology diffusion source. The transfer agency intermediary is generally aware of the technology, and owing to the specialization in the field of technology it serves, is generally familiar with the developments and applications in this field of technology. Therefore, this type of service institution not only performs the specific operation of implementing technical diffusion, but also provides consultation on technology advancement, applicability, and market application.

Innovation incubator-based intermediaries mainly provide production technology services and management services after innovation technology diffusion. The agency is committed to helping small and medium enterprises who adopt new technologies to raise and build the high-tech enterprises or projects, including supply credit, plant and machinery hire, staff recruitment, advisory services for production technology and management, including technical training for staff and vocational reeducation, new product identification, and quality certification, consultation for manufacturer, marketing information of raw material and products, and publicity and exhibition. The service supplied by innovation incubator-based intermediaries is conducive to overcome the obstacles in last two stages in technical diffusion.

The three types of intermediaries focus on specific areas, although the respective managements and services overlap sometimes. Various types of intermediary organizations combine to form an intermediary system, which can be properly used to eliminate the obstacles at different stages in the technical diffusion process.

Information intermediaries generally do not have the technical innovation, they merely serve as a means of communication of demand and supply contact information.

Innovation technology transfer agents serve innovative technology in a special field; they are familiar with the developments and applications in this technological field and can provide consultation for technology advancement, applicability, and market application.

Innovation incubator-based intermediaries mainly provide production technology and management services after technology diffusion. They can help build high-tech enterprises or projects, and provide production and management services.

All the factors which impact technical diffusion in the social environment of the guide-into region have promotional and restrictive functions on technical diffusion, but the relationship among these factors are very complex and ambiguous.

5.2.2 Resource Environment of Technical Diffusion Field

When considering the resources environment, reference is made not only to the local natural resources, but also the possibility of allocating resources in the society as a whole on the premise of improving economic efficiency as much as possible. The difference between the resource environments will result in different diffusion effects of the technology when spread in various regions.

5.2.2.1 Personnel Quality

The variability of human resources has a decisive influence on technical diffusion. Human resources are the most important part of the resource environment. Personnel quality is an important factor in technical diffusion, and to successfully proceed to technical diffusion, staff and production material must not only maintain a certain ratio, but also adapt to each other in quality.

Technical diffusion needs human intelligence. Whatever the method, the essence lies in the teaching and learning of technical knowledge. The biggest problem in the digestion and absorption of technology for developing countries is the low-quality of personnel, who do not have the necessary skills to transfer the technical data into actual production capacity. Therefore, to improve the efficiency of technical diffusion, the training and the introduction of talents must be strengthened and a favorable environment for personnel growth must be created.

Staff can generally be divided into four categories: leaders, managers, technicians, and production personnel.

For regional leaders and managers, their mental state, level of education, management capabilities, etc. are the primary factors affecting technical diffusion. The improvements in modern technology make the relationship among the factors affecting production more complex, rigorous, profound, and extensive in technical diffusion and production activities. Transferring technology to the production process not only requires close cooperation among workers, but more importantly, the organic integration of man with machine and science. Thus it requires leaders and managers with a good mental state, reasonable professional qualification, and high-level scientific management ability to deal correctly with all kinds of economic and technological issues and to configure them scientifically to obtain economic benefits.

The effect of technical diffusion can also be affected by the quality of scientific and technical personnel. For example, transferring technology to undeveloped regions is difficult; even for developed regions, the transfer of secret technology is very difficult. Low-quality personnel can only understand the detailed operation methods of the technology used in production, but will hardly be able to obtain any other information. Technical diffusion can only exert an effect on the unit to which technology has been transferred directly, but fail to promote technological progress across the region.

In the application of technology, more and more complex work requires a production staff at a high level of education and capability.

Staff management needs to pay attention to coordination ability, that is, from the system's angle, reasonably organize different types of personnel to enable an integrated function. At the same time, the quality training of potential personnel should be taken care of. Different educational structures bring up different specifications and personnel. The levels of national and science and technology education serve as measures of a country's capability level for digestion and absorption of advanced foreign technology.

5.2.2.2 Raw Materials

Raw material is one of the important elements of technology transfer. It is important to consider first, whether the local region can ensure the supply of raw materials and second, the proximity to the origin of raw materials to reduce the transportation of raw materials to the process of product manufacture, and hence, the costs of transport and labor.

Technical diffusion needs to consider supply capacity of the required raw materials in the application of technology. If the raw materials cannot be ensured, the high-speed development of the national economy as a result of technical diffusion will be limited considerably. Meanwhile, technical diffusion can also contribute to solving the problem of insufficient raw materials. Advanced diffusion can reduce the input of natural resources per unit product, and also create possibility for the integrated utilization of resources and the development of new materials.

5.2.2.3 Energy

First, it is necessary to ensure the supply of energy locally, second, ensure proximity to the origin of energy as much as possible. The details of energy production, transportation, and consumption should also be considered. Attention must also be given to the contribution of energy-saving technologies as a solution of the energy crisis.

5.2.2.4 Funds

Another important factor in technology diffusion is the availability of funds and their right use. Strong financial support could not only ensure the technology absorber will choose the most appropriate technology, but also be conducive to the technology import, promotion, improvement, and application.

5.2.2.5 Transportation and Communications

Transportation and communication affect many aspects, such as the size of the production reserve, the distance between the product and the market, capital turnover time, the time for commodity circulation, and the exploration of the international and domestic markets.

In technical diffusion, to achieve higher economic benefit, transportation must be given top priority and an effort must be made to reduce unreasonable transports and improve production efficiency. Meanwhile, the technology absorber should pay attention to the construction of roads, railways, ports, and other transportation facilities to provide good conditions for industrialization.

With the continuous development in science and technology, communication plays a more and more important role in economic development and intercommunication.

Good communication could enhance the relationship between the technology absorber and the source and be conducive to accomplish technical diffusion.

5.2.3 Technology Transfer Rate in the Technical Diffusion Field

Because of different development levels in society, in the economy, and in technology, as well as the different characteristics of natural resources, culture and value, the product levels of various regions are unbalanced.

The success standard of technical diffusion cannot be based on whether a local factory is well equipped or whether it produces under some concessions. The standard should consider whether this technology diffuses around the area and its speed and scope are the important evaluation criteria. In this book, this rate of speed together with the scope is recorded as the technology transfer rate.

5.2.3.1 Technical Relevance

Technical relevance refers to the following: first, the region has the related technological base before diffusion, second, it has promoted the progress of other local technologies after diffusion.

Technical relevance only reflects the actual use of the technology to some extent, in other words, transferred technology that is highly relevant to other departments also in the region. Actually a transferred technology can not only be used in the sector to which it is directed, but also promote the progress of technologies in other sectors through the effects of technical relevance.

The relevance of technology itself has a decisive impact on technical diffusion, different technologies with larger relevance have a stronger absorptive capacity among them and the speed of technical diffusion is high.

Because of differences in technology bases and structures in each area, the same technology has different relevance in different areas, thus affecting its diffusion rate in each area.

5.2.3.2 Applicability of Technology

This refers to the special conditions that the area, where the technology is to be introduced, must adapt to so that the effects are maximized.

"Applicability" is a relative concept which considers "who," "for whom," "how to use," "what kind of technology to introduce," etc. Different factors, like the state of production, social and resource environment, market conditions, and status of the technology, will undoubtedly lead to different levels of applicability of the same technology in different regions. If a technology has a higher applicability in the diverted areas, it will speed up the pace of diffusion in the region and achieve better results.

5.2.3.3 Technology Gap

The technology gap is the difference between the current levels of technology in a region and in the international scene.

The imbalance in the technological development level in each area has caused different gaps between different regions in the same technology. Only when there is a modest gap can the receiving region have both internal and external technical power so as to step up the speed of technology transfer and enhance the technology transfer rate.

A large gap between the level of transferred technology and the corresponding receiving region will affect the communication between them. Transferring by force will not only affect the direct goal, but also impede expansion and digestion of the technology and finally fail to achieve the aim of promoting the technology transfer. If the gap is too small, although the levels of technology among the regions are close to each other and communication and diffusion will be easy, yet the low effect on the progress in the technology in the diverted area make companies reluctant to introduce such technology. In this case also it fails to increase the technology transfer rate.

5.2.3.4 Investment per Unit Product

It is the average amount of investment spent on basic construction per unit product. The factors affecting investment include natural conditions such as engineering geology and hydrologic geology and whether local electricity, water supply, and transportation environment are available as needed. The basic requirements for technology transfer vary in different areas. To achieve the same productivity, investment in poor areas will be higher than in areas with good conditions, i.e. higher investment per unit product.

This is an index reflecting the effect of saving investment. The anticipated investment per unit product will directly affect the transfer of a technology and the diffusion speed as a result. For the same technology, the lower the investment per unit product, the faster it will diffuse.

5.2.3.5 Investment Recovery Period

The investment recovery period is the length of time calculated from the day that a technology is actually used in production until the time the profit equals the investment. So it is considered an important index to measure the investment profit. A short recovery time means a high profit and will undoubtedly motivate people to invest and accelerate the technology transfer rate.

5.2.3.6 Investment Risk

Investment risk is measured as a probability measure. Every investment decision of technology transfer in an enterprise will have various possibilities, with uncertain outcomes. This uncertainty is called investment risk. The primary reasons for investment risks in production are the uncertainties in the company's operation and sensitivity to external conditions beyond the company's control.

Because of the imbalance in society, in the economy, and in the technology in the diffused district, all the following factors such as technology relevance, technical adaptability, technical gap, investment per unit product, investment recovery period, and investment risk, affect the transferring rate. Technical relevance and technical adaptability are ascending functions of the technology transfer rate. Transfer rate will increase first and then decrease as the technical gap changes. If investment per unit product, investment recovery period, and investment risk are smaller than the optimum value, the technical transfer rate is an ascending function; when they exceed the optimum value, it becomes a decreasing function. The transfer rate is one of the significant factors deciding whether the region is the best area for technology transfer.

5.2.4 Power of Technical Diffusion

Technical diffusion and technical innovation are inseparable; in fact, technical diffusion is often accompanied by gradual technical innovation. Broadly speaking, the power of technical diffusion is the same as that of technical innovation.

5.2.4.1 Foreign Scholars' Opinions on the Power of Technical Diffusion

1. *The mechanism of the power of technical diffusion*: Broadly speaking, the concepts of the two mechanisms (technical diffusion mechanism and technical innovation mechanism) are the same. The scarcity of resources leads to the innovation. Hicks, a famous English economist, in his book *Wage Theory* published in 1932, argued that in the past several centuries, the main innovation related to labor-saving was in the direction of the change in relative prices of elements of production. "Factors of production, changes in relative prices, are in itself a particular incentive for invention—the invention lies in the direction of more economic use of those elements whose prices have become relatively expensive." The so-called relatively expensive elements were those becoming relatively scarce. As a result, two types of inventions happen, one, the "leading or causing invention"

and the other, "self-invention." In the direction of technical innovation, N. Rosenberg believed that induction mechanism was the moving force. But instead of being induced by scarcity factors, innovation derived from the following three mechanisms: the imbalance of technological development, production process, and resource supply uncertainties. What these three typical induced mechanisms had in common was that they were the toll-gates or bottlenecks of production, obstacles to the further development of production. Such obstacles created a pressure situation, which induced manufacturers to innovate to get around these barriers. Of course, every innovation was never permanent; it would create a new bottleneck, inducing people to reinnovate, and so the cycle would go on.

Japanese scholar Saito in the late 1970s proposed the assumption "N.R," which is primarily about the motivation for technical innovation. "N" represents "Needs" and "R" represents "Resources." For example, when there was a shortage of labor resources, there was a need for a labor-saving technical innovation; thus this factor played an important role in innovation. When other types of resources become insufficient, there would be a similar impact. The hypothesis of N–R relationship also includes the idea of technical innovation. In addition, when the resources cannot meet some kind of specified need, there would be a gap between "N" and "R"; This would serve as a stimulation factor to drive people to take action for technical innovation to close this gap. When technical innovation succeeds, the original gap between N and R is eliminated and the economy develops. And the measures taken to eliminate the gap are the development and researching activities in the process of technical innovations. In conclusion, technology need is the essential driving force for technical innovation by the researchers who undertake the task of technical development.

2. The power mode of technology diffusion
   a. *Technical promotion mode*: Schumpeter tends to believe the theory of the sovereignty of innovation in production which is closer to the technological promotion theory. From the fact that every country pays attention to science in their university education it is clear that until the 1960s most people believed the technological promotion theory of technology innovation.
   b. *Market attraction mode*: U.S. economist J. Schmookler concludes, based on the studies of patent activity, that patenting or invention activities, like other activities, were basically in the pursuit of economic profit which was guided and constrained by market demand.
   c. *Integrated action mode*: The integrated mode is generated on the basis of the balance between technical possibilities and market opportunities. N. Rosenberg believed that "innovation activities are decided by both the demand and technology where demand is in charge of the return of innovation and technology determines the possibility of success and the cost."

5.2.4.2 Domestic Scholars' Opinions of the Power of Technical Diffusion

Professor Liming Zhao, Tianjin University, in his book named "*Technology Transfer Theory*" argued that the "bottleneck" phenomenon between technology absorber and technology resources was the driving force behind technical diffusion. The motivation comes from both sides; lack of diffusion on either side would affect the process of diffusion. For the technical diffusion source, acceleration of technical diffusion can compensate the cost of technology development and can generate a satisfactory income. For the technology absorber, the adoption of technology can reduce the gap between the local and the developed region in terms of economy and technology.

In the book named "*Economics of Technological Progress*," professor Haishan Wang divided technical innovation into exogenous and endogenous power. Exogenous power comes into effect only through induction or when being transformed into endogenous power; and endogenous power makes use of the external "market power" as its source of energy in an orderly manner to generate an effective dynamic response and power circulation in the exterior environment, and to a larger extent promote technical innovation activities spontaneously. Exogenous power includes scientific progress, social needs, market competition, government policies, etc. Endogenous power includes the innovational consciousness of the subjects, the pursuit of the maximum economic benefits in enterprises or innovation institutions, and the inherent requirements of improving business competitiveness of enterprises.

Chunyou Wu, Dashuang Dai, and Jingqin Su, in the Dalian University of Technology, wrote in their book "*The Spread of Technological Innovation*," that the power or the driving force of technological innovation should take the technical innovation subjects as the core, including external and internal powers. External motivation is driven by technological invention and social needs. However there is only one internal motivation of technological innovation, namely the innovation subject's pursuit of maximum profit. Other factors should all be considered as the affecting elements of technological innovation rather than the power or the driving force of technological innovation.

5.3 Technical Diffusion Analysis of the Technical Diffusion Field

As technology is uneven in its spatial distribution, technology transfer exists not only between countries and regions, but also among industries and departments. As a result, technical diffusion is a multilevel and multi-dimensional diffusion.

5.3.1 Building of the Technical Diffusion Field

"Field" refers to the spatial distribution of a physical quantity. The physical quantity with a scalar characteristic has a scalar field in spatial distribution, whereas the other kind with a vector characteristic has a vector field.

At every point around a heat source such as a stove or radiator, temperature is certain, that is, there is some distribution of temperature in space called velocity field. At any point on earth, there is gravitational attraction on all objects, which is the gravity field. At various points around the charge, there exists a force, or electric field around the electric charge, etc. In the above-mentioned fields, the temperature field is a scalar field, whereas the electric field, velocity field, and gravity field are all vector fields.

Field is the spatial distribution of a physical quantity, which may change with time. Mathematically, a field is described by a multifunction, which employs the space and time coordinate to indicate its characteristic physical quantities. To be specific, the scalar field is a scalar function of space and time, and the vector field is a vector function of space and time. A reference coordinate system with main factors (technology transfer rate, social environment, and resource environment), which affect technical diffusion can be established and then the vector field can be analyzed in a quantitative way to demonstrate the space location and vector direction.

If the corresponding value of the physical quantity in every point of the field does not change with time, we call the field a steady or static field, otherwise as unstable or time-varying field. The position of an unstable field at any moment can be described by a steady field.

The technical diffusion field is a special form of material and information on this base. Technology bearers interact through the technical diffusion field.

5.3.2 Law of Technical Diffusion in the Technical Diffusion Field

In the branches of natural sciences and engineering technology, state variables vary not only with time but also with space. Take gas diffusion for example, in addition to change with time, the gas density is different at different places. In this case, it is not enough to only employ the model of time evolution, but also needs to consider relations in space location. As a result it can be solved within the framework of the dynamic model of space–time evolution, which is a partial differential equation.

The field, which obeys the law of causation, is the distribution of physical quantities. The cause is called the field source. A field is generated from the field source, even as the temperature field is derived from the heat source. The spatial distribution of a field not only depends on its source, but is also closely related to its surrounding physical environment. For example, the temperature distribution in the furnace is determined not only by the size and distribution of firepower, but also by its structure as well as the material. The relationship of field, source, and material can be described by a group of differential equations, for instance, the relationship of the electromagnetic field and its source conforms to a group of vector partial differential equations of Maxwell.

Chapter 6

Quantitative Study of the Technology Diffusion Field and Technology Dynamics

Technology diffusion essentially emanates from the production and reproduction of science and technology, invested directly into society, as the common wealth of human beings, science and technology diffuses among various countries, regions, fields, and the different classes, and performs as part of the productivity flow during their mutual communication, inheritance, and innovation process.

6.1 Quantitative Model of the Environment of the Technology Diffusion Field

6.1.1 *Environment of the Technology Diffusion Field*

Technology diffusion occurs in specific social and resource environments. The multivariable environments have complex interdependence and mutually restricting relations with technology diffusion as shown in Figure 6.1.

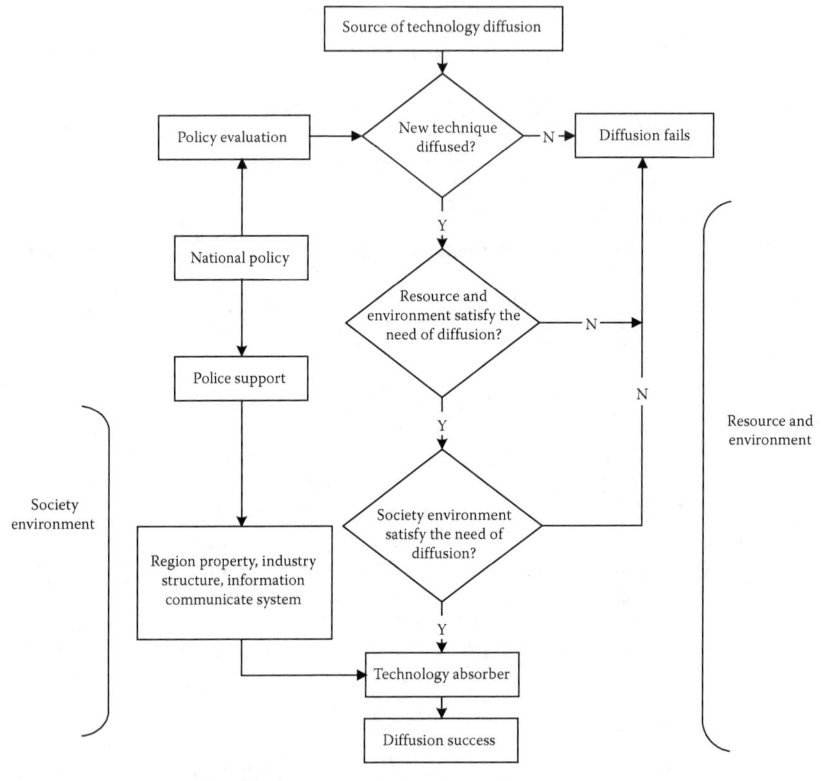

Figure 6.1 The relation between technology diffusion and environment.

6.1.2 Quantitative Model of the Environment of the Technology Diffusion Field

The environment of technology diffusion field $\mathbf{X}$ is a complex system, with the primary factor set $\mathbf{X}$ and subfactor set $\mathbf{X}_i$ $(i = 1, 2, \ldots, I)$. Figure 6.2 illustrates the model.

$$\mathbf{X}\text{--}\left|\begin{array}{l} \mathbf{X}_1\text{: } \mathbf{X}_{11}, \mathbf{X}_{12}, \mathbf{X}_{13}, \ldots, \mathbf{X}_{1k_1} \\ \mathbf{X}_2\text{: } \mathbf{X}_{21}, \mathbf{X}_{22}, \mathbf{X}_{23}, \ldots, \mathbf{X}_{2k_2} \\ \quad\ldots \qquad \ldots \qquad \ldots \\ \mathbf{X}_i\text{: } \mathbf{X}_{i1}, \mathbf{X}_{i2}, \mathbf{X}_{i3}, \ldots, \mathbf{X}_{ik_i} \end{array}\right|\text{--}q_m$$

$$(i = 1,2, \ldots, I;\ m = 1,2, \ldots, M)$$

$$(k_1 = 1,2, \ldots, K_1;\ k_2 = 1,2, \ldots, K_2;\ \ldots;\ k_i = 1,2, \ldots, K_i)$$

Figure 6.2 Model of the environment of technology diffusion field.

where

X is the integral environment of technology diffusion field

q_m is the evaluation target

$\mathbf{X}_1 - \mathbf{X}_i$ are the interweaving subsystems to constitute the integral environment

From Figure 6.2, we can see that:

$$\mathbf{X} = \{\mathbf{X}_i\}, \quad (i = 1, 2, \ldots, I) \tag{6.1}$$

Among them

$$\begin{aligned} \mathbf{X}_1 &= (\mathbf{X}_{11}, \mathbf{X}_{12}, \mathbf{X}_{13}, \ldots, \mathbf{X}_{1k_1}) \\ \mathbf{X}_2 &= (\mathbf{X}_{21}, \mathbf{X}_{22}, \mathbf{X}_{23}, \ldots, \mathbf{X}_{2k_2}) \\ &\cdots \quad \cdots \quad \cdots \\ \mathbf{X}_i &= (\mathbf{X}_{i1}, \mathbf{X}_{i2}, \mathbf{X}_{i3}, \ldots, \mathbf{X}_{ik_i}) \end{aligned} \tag{6.2}$$

6.1.3 Weight Calculation

6.1.3.1 1–6 Scale Method

1–6 scale method, as illustrated in Figure 6.3.

Suppose the comment set is as follows:

$$\begin{aligned} \mathbf{PYJ} &= \{\mathbf{PYJ}_j\} \\ &= \begin{Bmatrix} \text{Most important (Excellent), More important (Good), Important (Medium),} \\ \text{Normal (Bad), Less important (Inferior), Least important (Very inferior)} \end{Bmatrix}. \\ &= \{1.0 \quad 0.8 \quad 0.6 \quad 0.4 \quad 0.2 \quad 0\}, \quad (j = 1, 2, \ldots, 6) \end{aligned} \tag{6.3}$$

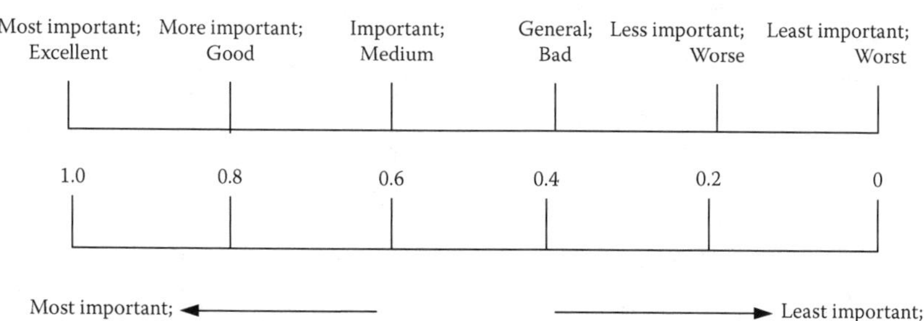

Figure 6.3 1–6 scale method.

6.1.3.2 Pairwise Comparison Matrix

First, arrange the indexes, $\mathbf{X}_{ik}$ (k = 1,2,…,K), of every subfactor set $\mathbf{X}_i$ (i = 1, 2,…,I) of the technology diffusion environment of the region being evaluated in the form of a matrix. Then invite 5–15 experts to give the importance of the corresponding index after a pairwise comparison of $\mathbf{X}_{ik}$. The comment set is shown as Formula 6.3. To be specific, find the average, denoted as $\mathbf{FZ}_{ik}$, of every index of the set $\mathbf{X}_i$ after the pairwise comparison of $\mathbf{X}_{ik}$ and $\mathbf{X}_{ik-1}$. Then sum up $\mathbf{FZ}_{ik}$, that is, $\mathbf{FZH}_i = \sum_{i=1}^{I}\sum_{k=1}^{K}\mathbf{FZ}_{ik}$. Finally, we will have the weight of every index of $\mathbf{X}_i$, namely $\mathbf{W}_i$, $\mathbf{W}_i$ = ($\mathbf{W}_{i1}$, $\mathbf{W}_{i2}$,…, $\mathbf{W}_{ik}$), and $\mathbf{W}_i$ = ($\mathbf{FZ}_{ik}$/$\mathbf{FZH}_i$). Table 6.1 illustrates the comparison matrix.

6.1.4 Fuzzy Evaluation Model of the Technology Diffusion Field

For the multilevel fuzzy evaluation model shown in Figure 6.2, the primary factor set is $\mathbf{X}$, the subfactor set is $\mathbf{X}_i$ (i = 1, 2,…, I), and the corresponding weight is

$$\mathbf{W} = (\mathbf{W}_i), \quad (i = 1,2,\ldots,I) \tag{6.4}$$

Among them

$$\begin{aligned}
\mathbf{W}_1 &= (w_{11}, w_{12}, w_{13}, \ldots, w_{1k_1}) \\
\mathbf{W}_2 &= (w_{21}, w_{22}, w_{23}, \ldots, w_{2k_2}) \\
&\cdots \quad \cdots \quad \cdots \\
\mathbf{W}_i &= (w_{i1}, w_{i2}, w_{i3}, \ldots, w_{ik_i})
\end{aligned}$$

Table 6.1 Pairwise Comparison Matrix

Region	$\mathbf{X}_{i2}$	$\mathbf{X}_{i3}$	…	$\mathbf{X}_{ik-1}$	$\mathbf{X}_{ik}$	$\mathbf{FZ}_{ik}$	$\mathbf{W}_i$ = ($\mathbf{FZ}_{ik}$/$\mathbf{FZH}_i$)
$\mathbf{X}_{i1}$			…				
$\mathbf{X}_{i2}$			…				
$\mathbf{X}_{i3}$			…				
…				…	…	…	…
$\mathbf{X}_{ik-1}$							
$\mathbf{X}_{ik}$							
Total						$\mathbf{FZH}_i$	1

$$(i=1,2,\ldots,I);(k_1=1,2,\ldots,K_1;\ k_2=1,2,\ldots,K_2;\ k_i=1,2,\ldots,K_i) \tag{6.5}$$

The weight and the comment set can be calculated according to Formulas 6.2 and 6.3 respectively.

The fuzzy relationship matrix of general object **R**, and the comprehensive evaluation subordinate degree **B** are as follows:

$$\mathbf{R}=\begin{vmatrix}\mathbf{R}_1\\ \mathbf{R}_2\\ \vdots \\ \mathbf{R}_i\end{vmatrix}=(r_{ij})_{i\times j},\quad (i=1,2,\ldots,I;\ j=1,2,\ldots,6) \tag{6.6}$$

$$\mathbf{B}=\mathbf{W}\cdot\mathbf{R}=(\mathbf{b}_j),\quad (j=1,2,\ldots,6) \tag{6.7}$$

r_{ij}, the average score decided by the experts according to Formula 6.3, demonstrates the degree of comment j when evaluating the target q_m in terms of the ith factor in set $\mathbf{X}_i$.

$$r_{ij}^{(2)}=\frac{r_{ij}}{n} \tag{6.8}$$

where n is the sum of every single row of r_{ij}.

$$\mathbf{B}_i=\mathbf{W}_i^{(2)}\cdot\mathbf{R}_i,\quad (i=1,2,\ldots,I) \tag{6.9}$$

$$\mathbf{R}_1=\begin{vmatrix} r_{11}^{(2)}, r_{12}^{(2)},\ldots,r_{1j}^{(2)}\\ r_{21}^{(2)}, r_{22}^{(2)},\ldots,r_{2j}^{(2)}\\ \cdots\quad\cdots\quad\cdots\\ r_{k_11}^{(2)}, r_{k_12}^{(2)},\ldots,r_{k_1j}^{(2)}\end{vmatrix}$$

$$\mathbf{R}_2=\begin{vmatrix} r_{11}^{(2)}, r_{12}^{(2)},\ldots,r_{1j}^{(2)}\\ r_{21}^{(2)}, r_{22}^{(2)},\ldots,r_{2j}^{(2)}\\ \cdots\quad\cdots\quad\cdots\\ r_{k_21}^{(2)}, r_{k_22}^{(2)},\ldots,r_{k_2j}^{(2)}\end{vmatrix} \tag{6.10}$$

$$\cdots\quad\cdots\quad\cdots$$

$$\mathbf{R}_i=\begin{vmatrix} r_{11}^{(2)}, r_{12}^{(2)},\ldots,r_{1j}^{(2)}\\ r_{21}^{(2)}, r_{22}^{(2)},\ldots,r_{2j}^{(2)}\\ \cdots\quad\cdots\quad\cdots\\ r_{k_i1}^{(2)}, r_{k_i2}^{(2)},\ldots,r_{k_ij}^{(2)}\end{vmatrix}$$

$$\mathbf{B}_i=\mathbf{W}_i\cdot\mathbf{R}_i=\bigvee_{i=1}^{I}\left(\mathbf{W}_i\wedge r_{ij}^{(2)}\right) \tag{6.11}$$

Suppose the fuzzy comprehensive comment set of "Standard Target **T**" is

$$\mathbf{U}_i = \{u_{ij}\}, \quad (j = 1,2,\ldots,6) \tag{6.12}$$

And the subordinate degree is

$$\hat{\mathbf{B}}_i = \mathbf{W}_i \cdot \mathbf{U}_i = \overset{I}{\underset{i=1}{\vee}} (\mathbf{W}_i \wedge u_{ij}) \tag{6.13}$$

According to the fuzzy-discrete closeness principle, we can calculate the closeness degree λ_i between **T** and the evaluation target **Q** as:

$$\lambda_i = \frac{1}{2}\left[\hat{\mathbf{B}}_i \cdot \mathbf{B}_i + (1 - \hat{\mathbf{B}}_i \odot \mathbf{B}_i)\right] \tag{6.14}$$

6.2 Quantization of the Social Environment of the Technology Diffusion Field

6.2.1 Quantitative Analysis of the Social Environment of the Technology Diffusion Field

1. Index system of the social environment of the technology diffusion field
   The social environment **X** is a complex system, whose index system is described in Figure 6.4 with the primary factor set **X** and subfactor system $\mathbf{X}_i$, (i = 1, 2,…, 8).
2. Quantitative calculation flowchart of the social environment of the technology diffusion field

Figure 6.5 shows the weight calculation flowchart. Figure 6.6 illustrates the subordinate degree and closeness degree of the technology diffusion environment; both can employ MATLAB® to obtain the results.

6.2.2 Establishment of Axis x of the Technology Diffusion Field

Let $\lambda_x = \lambda_i$,

$$x = \frac{1}{\lambda_x}, \quad (\lambda_x \neq 0) \tag{6.15}$$

It is beneficial for technology diffusion when $x \to \infty$.

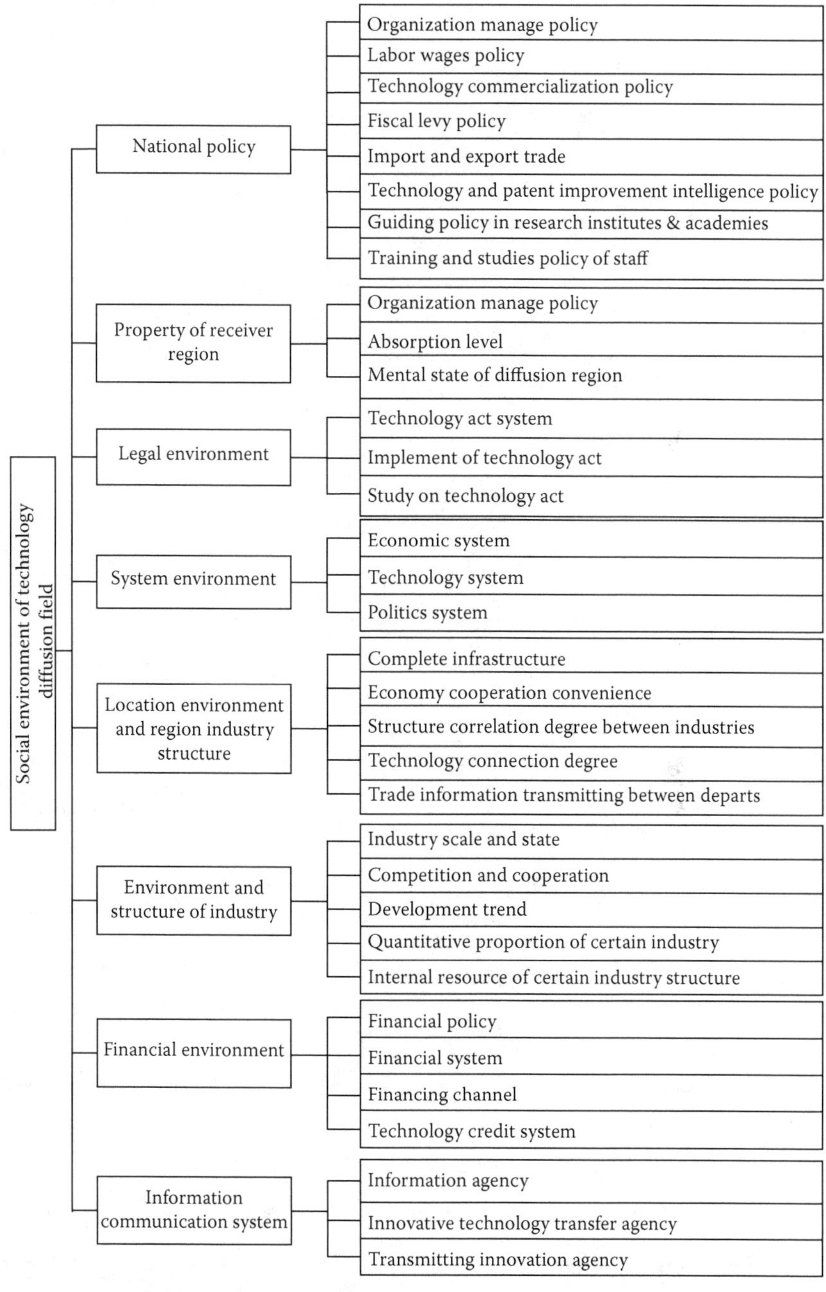

Figure 6.4 Index system on social environment of technology diffusion field.

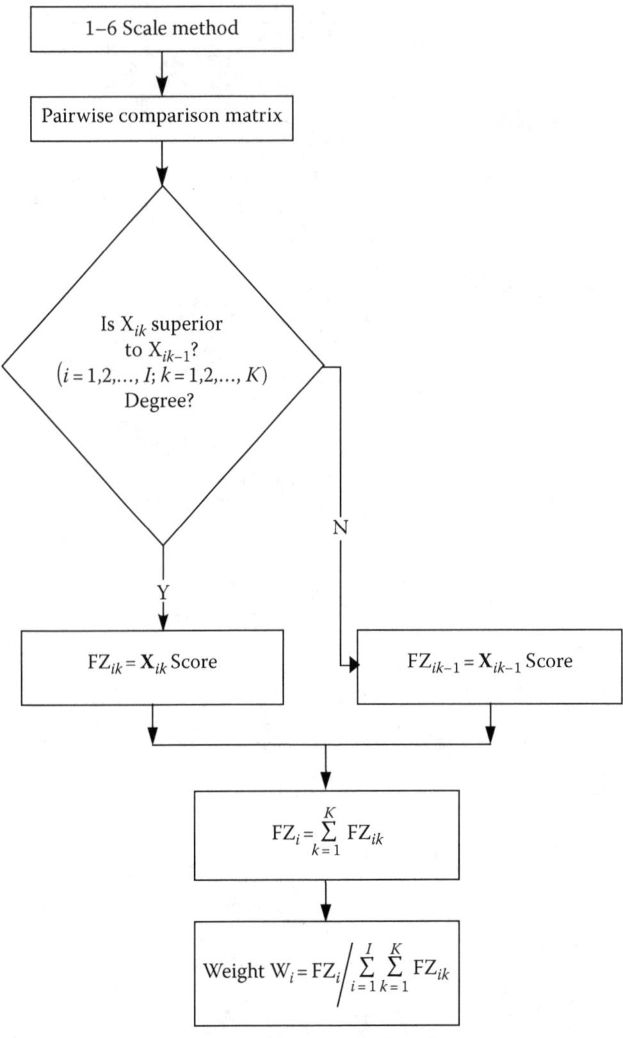

Figure 6.5 Weight calculation flowchart.

6.2.3 Simulation of Quantitative Calculation of the Social Environment of Technology Diffusion

1. Weights of the subfactor sets of the social environment of technology diffusion
   Let $\mathbf{X}$ and $\mathbf{X}_i$ be the primary factor and subfactor set of the technology diffusion social environment in target region q_m respectively.

$$i=1,2,\ldots,8;\quad m=1;\quad k_1=1,2,\ldots,8;\quad k_2=1,2,3;\quad k_3=1,2,3;$$

$$k_4=1,2,3;\quad k_5=1,2,\ldots,5;\quad k_6=1,2,\ldots,5;\quad k_7=1,2,\ldots,4;\quad k_8=1,2,3$$

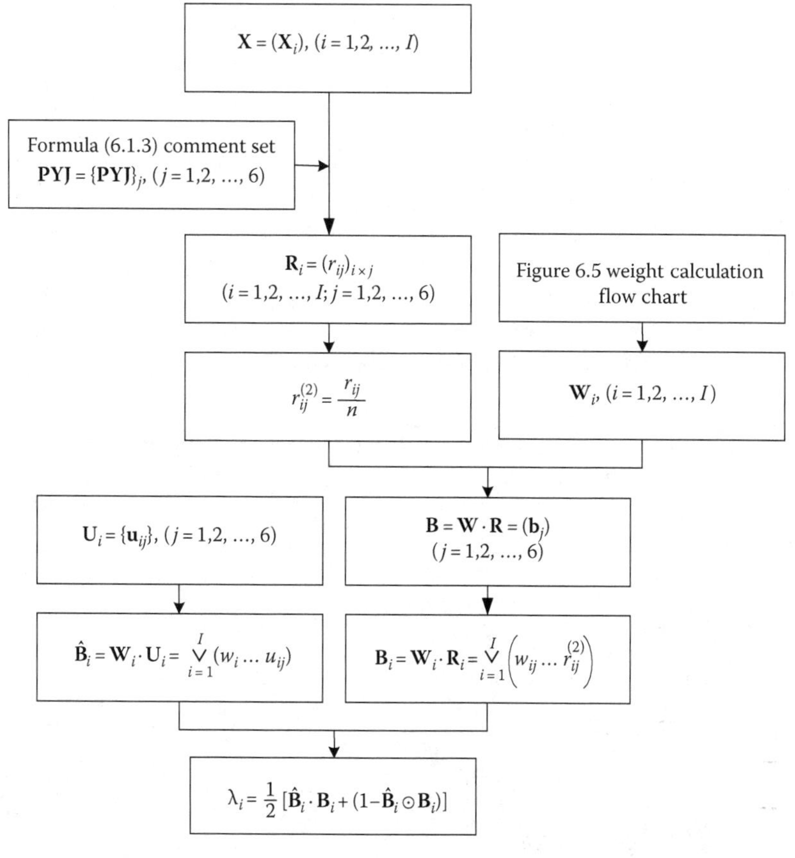

Figure 6.6 Calculation flowchart of subordinate and closeness degree of the environment of technology diffusion field.

From Figure 6.4, we can get the environment model of the technology diffusion field, as in Figure 6.7.

Suppose that 10 experts in this area successively compare pairwise the index of every subfactor set $\mathbf{X}_i$ of the social environment according to the comment set shown in Formula 6.3. Then we can get the significance of the corresponding index.

The simulation of the degree of significance according to each expert is to employ random numbers in the importance comment set, shown in Formula 6.3. Then MATLAB is used, following the calculation flowchart. According to Figure 6.5, we can calculate the weight by MATLAB program.

The results of the subfactor set weights in the social environment of the technology diffusion field are illustrated in Table 6.2.

$$X--\left|\begin{array}{l}X_1: X_{11}, X_{12}, X_{13}, X_{14}, X_{15}, X_{16}, X_{17}, X_{18}\\X_2: X_{21}, X_{22}, X_{23}\\X_3: X_{31}, X_{32}, X_{33}\\X_4: X_{41}, X_{42}, X_{43}\\X_5: X_{51}, X_{52}, X_{53}, X_{54}, X_{55}\\X_6: X_{61}, X_{62}, X_{63}, X_{64}, X_{65}\\X_7: X_{71}, X_{72}, X_{73}, X_{74}\\X_8: X_{81}, X_{82}, X_{83}\end{array}\right|--q_1$$

Figure 6.7 Environment model of the technology diffusion field.

Table 6.2 Weights of the Subfactor Set in the Social Environment of Technology Diffusion Field

w_{ik}		*k*							
		1	2	3	4	5	6	7	8
i	1	0.17916	0.12759	0.13067	0.11500	0.17136	0.09400	0.07483	0.10741
	2	0.13048	0.53181	0.33770					
	3	0.38944	0.27833	0.33222					
	4	0.22460	0.46528	0.31012					
	5	0.24882	0.19380	0.18495	0.17931	0.19312			
	6	0.17144	0.29222	0.17217	0.15662	0.20755			
	7	0.31326	0.17524	0.13141	0.38009				
	8	0.24964	0.21286	0.53750					

2. Subordinate degree and closeness of the subfactor set in the social environment of the technology diffusion field
   a. Calculation of set subordinate degree

      r_{ij}, the average of the score given by the ten experts according to Formula 6.3, demonstrates the degree of comment *j* when evaluating the target q_m in terms of the *i*th factor in set $\mathbf{X}_i$.

      Random numbers are employed to simulate the degree of importance as judged by experts according to the significance comment set of Formula 6.3. They are calculated in a row-normalized way and finally the average is obtained. Suppose the weight value is as Table 6.2 shows, then MATLAB is used to calculate as in Figure 6.5 and the program is shown in Appendix B, as the program of the environment subordinate and closeness degree of the technology diffusion field.

The results of the set subfactor subordinate degree in the social environment of technology diffusion field are illustrated in Table 6.3.

Let the comment set of the fuzzy relationship matrix of the subfactor class in **T** be

$$\mathbf{U}_i = \{u_{ij}\} = \{\text{Excellent}, 0, 0, 0, 0, 0\} \tag{6.16}$$

The weight is as shown in Table 6.2.

MATLAB is used according to Figure 6.6 to find the subordinate degree of the subfactor class of **T**, as shown in Table 6.4; the program is illustrated in Appendix B.

b. Calculation of closeness degree
From Formula 6.1.13, the closeness degree between the subfactor set of the social environment and the subfactor layer of **T** can be obtained through MATLAB. The closeness degree of their corresponding subordinate degrees is demonstrated in Table 6.5.

3. Weight, subordinate, and closeness degree of the primary factor set of the social environment of the technology diffusion field
Suppose in target region q_m, the primary factor set of the social environment of technology diffusion field is **X** and the subfactor set is $\mathbf{X}_i$.

$$m = 1; \quad i = 8$$

Table 6.3 Subfactor Set of the Subordinate Degree of the Social Environment of Technology Diffusion Field

B_{ij}		*j* = 1	2	3	4	5	6
i	1	0.08431	0.15181	0.06128	0.08736	0.12929	0.09821
	2	0.13377	0.10131	0.20636	0.10000	0.11305	0.27260
	3	0.03894	0.16139	0.10000	0.17217	0.17756	0.17217
	4	0.13958	0.20000	0.12940	0.21391	0.15508	0.10855
	5	0.14256	0.12989	0.12721	0.17788	0.09361	0.06908
	6	0.12915	0.12724	0.12597	0.18068	0.14130	0.03281
	7	0.10438	0.15553	0.19628	0.11818	0.09770	0.15204
	8	0.15007	0.04625	0.24257	0.17504	0.21118	0.10368

Table 6.4 Subordinate Degree of the Subfactor Layer of T

$\hat{B}_{ij}$		*j*					
		1	2	3	4	5	6
i	1	0.17916	0	0	0	0	0
	2	0.53182	0	0	0	0	0
	3	0.38944	0	0	0	0	0
	4	0.46528	0	0	0	0	0
	5	0.24882	0	0	0	0	0
	6	0.29222	0	0	0	0	0
	7	0.38009	0	0	0	0	0
	8	0.53750	0	0	0	0	0

Table 6.5 Subfactor Set of the Closeness Degree of the Social Environment of Technology Diffusion Field

i	1	2	3	4	5	6	7	8
λ_i	0.53640	0.56688	0.51947	0.56979	0.57128	0.56457	0.55219	0.57504

Table 6.6 Primary Factor Weight of the Social Environment of Technology Diffusion Field

i	1	2	3	4	5	6	7	8
W_i	0.11343	0.17678	0.11079	0.10849	0.08730	0.13751	0.13785	0.12785

Repeat step (1) in Section 6.2.3 to get the weight of the primary factor of the social environment of the technology diffusion field shown in Table 6.6.

Let $r_{i1} = \lambda_i$, and the weight as in Table 6.6. Step (2) in Section 6.2.3 is repeated and the primary factor of the subordinate degree of the social environment is obtained, as illustrated in Table 6.7.

Suppose the comment set of the fuzzy relation matrix of the primary layer of the Standard Object is as shown in Formula 6.16, and the weight as in Table 6.6. Step (2) in Section 6.2.3 is repeated to calculate the subordinate degree of its primary factor layer as shown in Table 6.8.

Table 6.7 Primary Factor Subordinate Degree of the Social Environment of Technology Diffusion Field

i	1	2	3	4	5	6
B_i	0.06176	0.07532	0.10506	0.08477	0.10776	0.04016

Table 6.8 Subordinate Degree of the Primary Factor Layer of T

i	1	2	3	4	5	6
$\hat{B}_i$	0.17678	0	0	0	0	0

According to Formula 6.2.13, the closeness degree between the primary factor set of the social environment and the primary factor **T** can be obtained. As obtained by MATLAB following Figure 6.6, the closeness degree of their subordinate degrees is:

$$\lambda_x = \lambda_i = 0.52652$$

That is,

$$x = \frac{1}{\lambda_x} = 1.89926$$

6.3 Quantization of the Resource Environment of the Technology Diffusion Field

6.3.1 Quantitative Analysis of the Resource Environment of the Technology Diffusion Field

1. Quantitative index system
   The resource environment **Y** is a complicated system with the primary factor set **Y** and subfactor set $\mathbf{Y}_i$, ($i = 1, 2, \ldots, 5$). The index system is shown in Figure 6.8.
2. Calculation flowchart of the quantization of resource environment
   Figure 6.5 shows the weight calculation flowchart and Figure 6.6 the subordinate and closeness degree calculation flowchart. Both can be created using MATLAB.

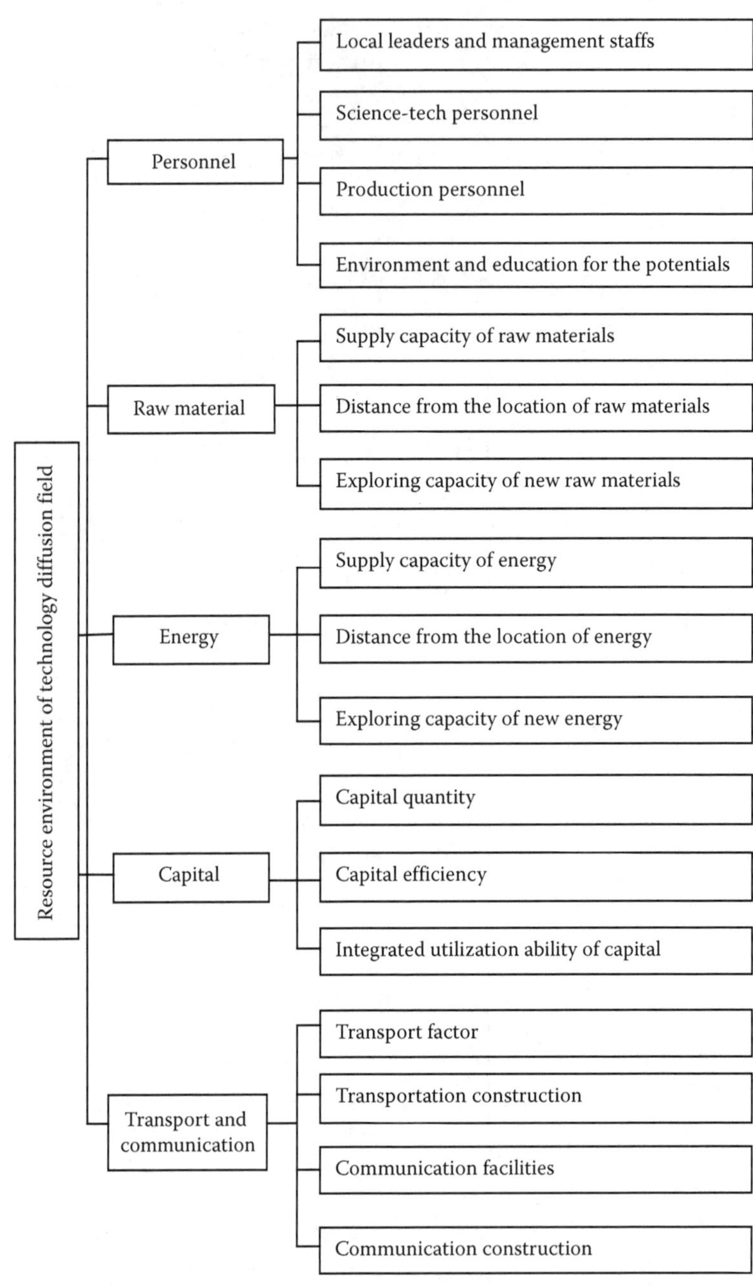

Figure 6.8 Quantitative index system of the resource environment of technology diffusion field.

6.3.2 Establishment of Axis y of the Technology Diffusion Field

Let $\lambda_y = \lambda_i$,

$$y = \frac{1}{\lambda_y}, \quad (\lambda_y \neq 0) \tag{6.17}$$

It is beneficial to technology diffusion when $y \to \infty$.

6.3.3 Simulation Calculation of the Resource Environment Quantization of the Technology Diffusion Field

1. The weight of the subfactor set

   In the evaluated region q_m, let **Y** be the primary factor and $\mathbf{Y}_i$ be the subfactor set of the resource environment.

   $$i = 1,2,\ldots,5; \quad m = 1;$$

   $$k_1 = 1,2,3,4; \quad k_2 = 1,2,3; \quad k_3 = 1,2,3; \quad k_4 = 1,2,3; \quad k_5 = 1,2,3,4$$

   The environment model of the technology diffusion field can be obtained according to Figure 6.8, as shown in Figure 6.9.

   Step (1) in Section 6.2.3 is repeated to get the weights of the subfactor set of the resource environment of the technology diffusion field, as shown in Table 6.9.

2. Subordinate and closeness degree of the subfactor set in the resource environment of the technology diffusion field

   Suppose the comment set for the fuzzy relation matrix of the standard object subfactor layer is as given in Formula 6.16, and the weights as shown in Table 6.9.

   Repeating step (2) in Section 6.2.3, gives the subordinate degree of the subfactor set of the resource environment as shown in Table 6.10. Table 6.11 illustrates the subordinate degree of the **T** subfactor set while Table 6.12 shows the closeness degree of the subfactor set of the resource environment of the technology diffusion field.

$$\mathbf{Y}\text{--}\left|\begin{array}{l} \mathbf{Y}_1: \mathbf{Y}_{11}, \mathbf{Y}_{12}, \mathbf{Y}_{13}, \mathbf{Y}_{14} \\ \mathbf{Y}_2: \mathbf{Y}_{21}, \mathbf{Y}_{22}, \mathbf{Y}_{23} \\ \mathbf{Y}_3: \mathbf{Y}_{31}, \mathbf{Y}_{32}, \mathbf{Y}_{33} \\ \mathbf{Y}_4: \mathbf{Y}_{41}, \mathbf{Y}_{42}, \mathbf{Y}_{43} \\ \mathbf{Y}_5: \mathbf{Y}_{51}, \mathbf{Y}_{52}, \mathbf{Y}_{53}, \mathbf{Y}_{54} \end{array}\right|\text{--}\mathbf{q}_1$$

Figure 6.9 Environment model of the technology diffusion field.

3. Weight and closeness degree of the primary factor set

   In the evaluated region q_m, let **Y** be the primary factor and $\mathbf{Y}_i$ be the subfactor set of the resource environment in the technology diffusion field.

Table 6.9 Weights of the Subfactor Set of the Resource Environment of Technology Diffusion Field

w_{ik}		k			
		1	2	3	4
i	1	0.29311	0.17176	0.28748	0.24766
	2	0.33715	0.27833	0.38452	
	3	0.29187	0.30730	0.40083	
	4	0.44000	0.23750	0.32250	
	5	0.29603	0.30624	0.20860	0.18913

Table 6.10 Subordinate Degree of the Subfactor Set of the Resource Environment

B_{ij}		j					
		1	2	3	4	5	6
i	1	0.05806	0.20853	0.08843	0.13273	0.18282	0.10056
	2	0.10000	0.20114	0.13845	0.19412	0.18279	0.06629
	3	0.05992	0.27081	0.18171	0.15992	0.16927	0.12919
	4	0.04400	0.19650	0.17975	0.17950	0.24425	0.04400
	5	0.10874	0.16743	0.09218	0.15006	0.19382	0.10102

Table 6.11 Subfactor Set Subordinate Degree of Standard Object T

B_{ij}		j					
		1	2	3	4	5	6
i	1	0.29311	0	0	0	0	0
	2	0.38452	0	0	0	0	0
	3	0.40083	0	0	0	0	0
	4	0.44000	0	0	0	0	0
	5	0.30624	0	0	0	0	0

Table 6.12 Closeness Degree of the Subfactor Set of Resource Environment of Technology Diffusion Field

i	1	2	3	4	5
λ_i	0.52903	0.55000	0.52996	0.52200	0.55437

Table 6.13 Weights of the Primary Factor of the Resource Environment

i	1	2	3	4	5
W_i	0.28562	0.30139	0.07106	0.14938	0.19254

Table 6.14 Subordinate Degree of the Primary Factor of the Resource Environment

i	1	2	3	4	5	6
B_i	0.11425	0.05492	0.17023	0.16090	0.21345	0.05272

Table 6.15 Subordinate Degree of the Primary Factor Set of Standard Object T

i	1	2	3	4	5	6
$\hat{B}_i$	0.30139	0	0	0	0	0

$$m = 1; \quad i = 5$$

By repeating step (1) in Section 6.2.3 the subfactor weights of the primary factor of the resource environment as shown in Table 6.13 can be got.

Let $r_{i1} = \lambda_i$, and the weights as shown in Table 6.13. By repeating step (2) in Section 6.2.3 the subordinate degree of the primary factor of the resource environment as shown in Table 6.14 can be got.

Suppose the comment set for the fuzzy relationship matrix of the subfactor layer of standard object **T** is as Formula 6.16 demonstrates, and the weights are as shown in Table 6.13. Step (2) in Section 6.2.3 is repeated to get the subordinate degree of the primary factor set of **T** as shown in Table 6.15.

According to Formula 6.1.14, the closeness degree between the subordinate degrees of the social environment primary factor set and the subject primary factor can be obtained by using MATLAB. According to Figure 6.6, the closeness degree of their subordinate degrees is:

$$\lambda_y = \lambda_i = 0.55712$$

That is,

$$y = \frac{1}{\lambda_y} = 1.79495$$

6.4 Technology Transfer Rate of the Technology Diffusion Field

6.4.1 Technology Transfer Rate

The general formula of Cobb–Douglas Production function is

$$Y = AK^{\alpha}L^{\beta} \tag{6.18}$$

where

Y is the output (gross output)
K is the capital
L is the labor (number of workers)

Elasticity of capital: $\alpha = \frac{\partial Y}{\partial K} \cdot \frac{K}{Y}$

Elasticity of labor: $\beta = \frac{\partial Y}{\partial L} \cdot \frac{L}{Y}$

$\alpha + \beta = 1$
where A is the production efficiency parameter (average product capacity here)
We call L^{α} and K^{β} production elements, denoted as M.

Then Formula 6.18 is

$$Y = AM \tag{6.19}$$

where $Y = \dot{Y}(t)$; $A = A(t)$; $M = M(t)$.

The change or increase in Y caused by the change in t is denoted by ΔY

$$\Delta Y \approx dY = MdA + AdM \tag{6.20}$$

Thus the increased rate of output is

$$\frac{\Delta Y}{Y} \approx \tilde{Y} = \frac{dY}{Y} = \frac{MdA + AdM}{Y} \tag{6.21}$$

Substituting Formula 6.21 for Formula 6.19

$$\tilde{Y} = \frac{dA}{A} + \frac{dM}{M} \tag{6.22}$$

Let

$$\tilde{A} = \frac{dA}{A} \tag{6.23}$$

$$\tilde{M} = \frac{dM}{M} \tag{6.24}$$

Formula 6.22 is

$$\tilde{Y} = \tilde{A} + \tilde{M} \tag{6.25}$$

It means the increased rate of output equals the sum of the increased rates of production capacity and production elements approximately.

Parameter A represents the average production capacity of the technology level in one region. A^* represents the new production capacity with a higher technology level, when applying technologies in all available areas.

Theoretically, thanks to the application of new technology, the transfer of production capacity is $A - A^*$, which is not true in fact. Although new technology can raise production capacity, they are not necessarily equal. Only when applied in every possible production area, and found in all productivity factors, can it equal the theoretical A^*, with the increase reaching $A - A^*$.

It is difficult for all to adopt this new technology in a short time, because of the features of the technology itself and regional discrepancy. When many factors restrict technology diffusion, the users' technology transfer rate is denoted as γ.

After introducing new technologies the regional production increases by ΔA, then

$$\Delta A = \gamma(A^* - A), \quad (A^* \geq A; 0 \leq \gamma \leq 1) \tag{6.26}$$

6.4.2 Quantitative Analysis of the Technology Transfer Rate of the Technology Diffusion Field

1. Quantitative index system for the factors influencing technology transfer rate
   Technology transfer rate γ is a complicated system. The quantitative index system for the factors influencing it is shown in Figure 6.10 with the primary factor set γ and subfactor set γ_i, $(i = 1, 2, \ldots, 6)$.
2. Quantitative calculation flowchart of the technology transfer rate of the technology diffusion field
   Step 6.1.4 is repeated to find out how many of these factors can influence the technology transfer rate. Figure 6.5 shows the flowchart of weight calculation.
   Let the subordinate degree of "standard object **T**" be

$$\hat{\mathbf{B}}_i = (1,0,0,0,0,0) \tag{6.27}$$

   Step 6.1.5 is repeated to calculate the subordinate degree λ_i of the technology transfer rate. The MATLAB program can be used for calculation.

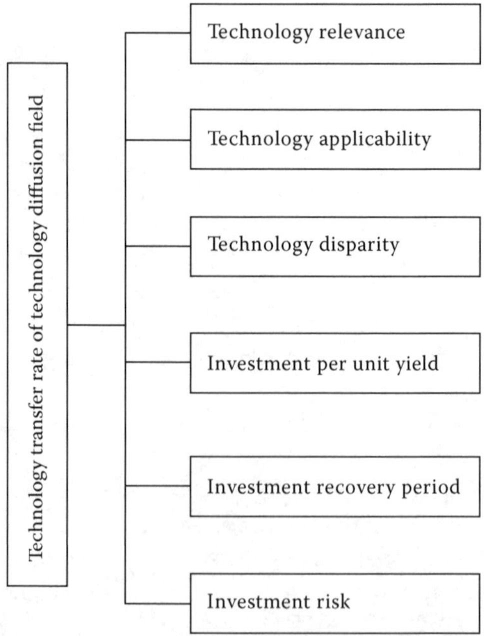

Figure 6.10 Quantitative index systems for the influencing factors of technology transfer rate.

6.4.3 Establishment of Axis z of the Technology Diffusion Field

The growth rate of productivity is

$$\tilde{A} = \frac{\Delta A}{A} \tag{6.28}$$

Substituting Formula 6.28 for Formula 6.26

$$\tilde{A} = \frac{\gamma(A^* - A)}{A} = \gamma\left(\frac{A^* - A}{A}\right) \tag{6.29}$$

Suppose the production factor M remains unchanged. To obtain the maximum yield growth rate, A^* should get its maximal value. However even in this case, yield growth rate could still be zero, that is, $\gamma = 0$, because of the lack of ability to absorb this technology in the absorption region. In conclusion, technology transfer rate decides the effect of this activity. When all adopt the new technology in the region, $\gamma = 1$, whereas on the contrary $\gamma = 0$ for social, economic, or technological reasons.

Let $\lambda_\gamma = \lambda_i$

$$z = \frac{1}{\lambda_\gamma}, \quad (\lambda_\gamma \neq 0) \tag{6.30}$$

It is beneficial for technology diffusion when $z \to \infty$.

6.4.4 Simulation of the Quantization of the Technology Transfer Rate

1. Subfactor weights of the primary factor of technology transfer rate of the technology diffusion field.

   Suppose the primary factor of technology transfer rate in the evaluated region q_m is γ, and its subfactor set is γ_i

   $$i = 1; \quad m = 1; \quad k_1 = 1, 2, \ldots, 6;$$

   From Figure 6.9 we can get the environment model of the technology diffusion field, as shown in Figure 6.11.

   Step (1) in Section 6.2.3 is repeated to calculate the subfactor weight constituting the primary factor of technology transfer rate, as illustrated in Table 6.16.

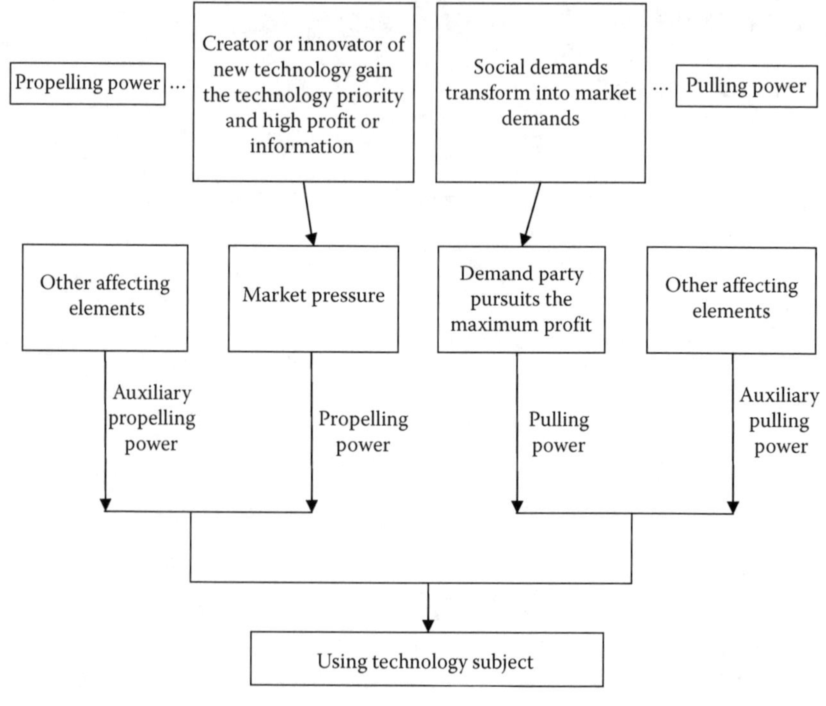

Figure 6.11 Environment model of the technology diffusion field.

Table 6.16 Primary Factor Weights of Technology Transfer Rate

i	1	2	3	4	5	6
W_i	0.16157	0.25282	0.05561	0.20863	0.18263	0.13876

2. Subordinate degree of the primary factor of technology transfer rate
   Suppose that the subordinate degree of "standard object **T**" is as illustrated in Formula 6.27 and the weight is as shown in Table 6.16.

   Step 6.2.4.2 is repeated to get the subordinate degree of technology transfer rate, shown in Table 6.17. The closeness degree of the subordinate degrees of technology transfer rate and **T** is $\lambda_\gamma = \lambda_i = 0.52660$.

   That is, $z = \dfrac{1}{\lambda_\gamma} = 1.89897$.

   The space coordinate is established using the axis of λ_x, λ_y, λ_γ to form the integral environment of the technology diffusion field. For research convenience, the technology source is taken as the origin and the axis of x, y, z is

Table 6.17 Subordinate Degrees of Technology Transfer Rate

i	1	2	3	4	5	6
B_i	0.05321	0.16355	0.16843	0.08270	0.13830	0.06444

employed to construct a three dimensional space coordinate. The technology source evokes the diffusion field and any absorber A in the field owning its coordinate $A(x, y, z)$. The distance between the diffusion origin O and A is

$$r = |\overline{OA}| = \sqrt{(x-x_0)^2 + (y-y_0)^2 + (z-z_0)^2}, \quad (x, y, z \in R^3) \tag{6.31}$$

r is the technology diffusion environment where the source and absorber belong.

Suppose the technology diffusion source to be the origin and the best environment factor is O $(0,0,0)$. The bigger $|x|,|y|,|z|$ is, the worst it is for technology diffusion and absorption. We can use this distance to represent the influencing factors of technology diffusion abstractly. The smaller the r, the better the environment is for technology diffusion, and vice versa.

$$r = |\overline{OA}| = \sqrt{x^2 + y^2 + z^2}, \quad (x, y, z \in R^3) \tag{6.32}$$

6.5 Quantitative Analysis of the Diffusion Dynamic of the Technology Diffusion Field

6.5.1 Diffusion Dynamic of the Technology Diffusion Source and Technology Absorber

1. *Diffusion dynamic of the technology diffusion source*: Despite the trends of security and monopoly, the technology diffusion source is forced to disperse due to interior and exterior factors. Accelerating this diffusion process can compensate the deployment cost and provide a satisfying profit. With technology being upgraded continuously, technology sources will lose much of the profit unless they make the best use of the opportunities. Meanwhile, the various owners of the same advanced technology will compete in this diffusion process, and catalyze it.
2. *Diffusion dynamic of the technology absorber*: The technology absorber can provide the following influences and benefits and lead to technology diffusion.

a. *Optimizing the local industry structure*: It will support and develop the leading local industries, spread new technologies to promote development and drive other industries that contribute to the national economy. A regional industry system having the features of high efficiency and coordinated increase, which meets the demands of economic development, will be established.
b. *Achieving sequence breakthrough in industry*: Industries will achieve sequence breakthrough under the new technology. The prospective, retrospective, and existing influences of the emerging departments will help to establish an efficient system that meets the demands of regional economic development. The structure of enterprises in an industry is altered to provide economic benefits and allow the absorption of diffusion technologies to distribute production on a scientific basis. As a result, it can effectively minimize waste and optimize the distribution process.
c. *Bridging the gap between economy and technology in developed regions to enhance competitive capacity*: For the technology absorber, the diffusion can accelerate technology transformation in enterprises and improve its research and development ability to achieve a cycle of input, digestion, innovation, and output. Yet we have to admit that increasing the technology level may not necessarily bring visible effects. Therefore it should take the local level into consideration to ensure that technology transformation really works and becomes the driving force of technology diffusion.
d. *Unifying the economic and social benefits*: Successful technology diffusion can increase economic benefits. A high level of technology will minimize the waste of resources by paying the least or by obtaining maximum value with the same amount of labor and resources.

   One important criterion to evaluate social and economic benefits is meeting social requirement. Enhancing economic benefits will undoubtedly resolve the conflict between demand and supply. Meanwhile the potential requirement should not be overshadowed by a short-time unilateral demand for short-term economic benefits.
e. *Enlarging the labor force*: It can expand production and reduce social differences. With the appearance of new materials, further exploitation of resources and emerging production departments, the scope for employing labor will be greatly enlarged.
f. *Great attraction of the vast local market*: Industries located near the consumption area improve productivity. Accelerating the turnover speed of the circulating capital and catering to consumers, play an important role.
g. *Promoting sustainable development*: Sustainable development is to seek a way that gives consideration to both the interests of contemporary people

and their descendants, and to realize coordination in society, the economy, the population, the resources, and the environment as well as to give sufficient attention to the ecologic deficit and environment overdraft in the process of industrialization and urbanization. Meanwhile, once the sustainable development strategy has been implemented, it will generate huge benefits in terms of reasonable use of resources, industry optimization, and ecologic environment protection, thus improving the quality of economic increase and promoting its healthy development, and finally laying the foundation to uplift the comprehensive national power.

Currently, environment pollution breaks the ecological balance, damaging and threatening human security. The main causes for this are the industrial pollutants, which are essentially because of the loss and waste of resources and energies. Through technology diffusion, new technologies, techniques, materials, and equipments, which cause no or little pollution will be adopted to make full use of resources and energies and to decrease the pollutants to a minimum.

h. Solving the problems of poor technology equipments and uncomfortable working environment in factories.

In some enterprises, poor technology equipments and an uncomfortable working environment could cause accidents and professional diseases. Technology diffusion can liberate workers from these dangerous conditions and protect their health and working abilities.

We argue that the power of technology diffusion comprises propelling and pulling forces. In the early period of technology creation and innovation those who adopted the new technical achievements gained excessive profits, thus creating the market competition pressure called the propelling force. The pulling force is their pursuit of maximum profit. The joint action of these two powers promotes the diffusion of the new technical achievements. The relation between the technology diffusion powers is shown in Figure 6.12.

There are many advantages brought about by the new technical achievements, for instance, meeting market demands, better satisfying the present requirements, achieving unprecedented high efficiency, and saving a large amount of social resources thanks to low consumption. It will also bring excessive profit to the creator or innovator or those adopting the new achievements in the early years

$$\gamma -- \left| \gamma_1 : \gamma_{11}, \gamma_{12}, \gamma_{13}, \gamma_{14}, \gamma_{15}, \gamma_{16} \right| -- \mathbf{q}_1$$

Figure 6.12 Relation of technology diffusion powers.

and raise their competitive ability. This market competition pressure propels other enterprises to adopt the new technology. In a word, the invention and innovation of technology is the propelling force of technology diffusion. The early users of these achievements retain their technical superiority and high profits thus creating a market for competition, which promotes technology diffusion.

If later users of the new technical achievements cannot win the market or surpass their original profit level, then they would have no desire to adopt this new technology. For users, the mere existence of social demands does not necessarily result in action; only a pulling source, which will be transformed into real marketing demands when the appearance of the new technology conforms to people's requirement, their ability to pay, and overcomes the barriers to their ideals and consciousness will count. Those who adopt the technology win a larger quota in the market and excess profit, and the desire to adopt the new technology is born in others.

The exterior driving power of the new technology users is the power of using this achievement and the interior power is the propelling power generated by increasing profit of the later users because of market demands. The two powers working at the same time only can promote technology diffusion.

The essence of power is to win profit and the ability to gain excessive profit is provided by the technology process or technology level. Scientific discovery and technology invention create new social demands based on the original one that the technology diffusion sources rely on to conduct the technology invention and innovation and spread the technology later. The basic impetus of this activity is the excessive profit brought about by adopting new technical achievements. And the technology absorber accepts them to satisfy social requirements, which is also driven primarily by a desire to maximize profits. The ability for this activity is provided by the technology progress or technology level. For the sake of profit, technology diffusion source and absorber should maintain the process of technology invention and innovation, diffusion, and absorption. During the period of operation of these phases, the propelling power is the desire of the technology source for excessive profits, and the pulling power is the impetus of the technology absorber for maximum profits. Technology progress is the capacity for both of these powers. Further more, other factors and auxiliary pulling powers are also integrated into the power of technology diffusion besides the two listed above.

6.5.2 Quantization of the Dynamic of Technology Diffusion Field

Cobb–Douglas product function and its modifying and derivative model

In 1928, the American mathematician C.W. Cobb and American economist P.H. Douglass obtained the macro production function $Q = 1.01K^{0.75}L^{0.2}$ based on America's statistical data from 1899 to 1922. Later the function was changed to Formula 6.33 to study the micro production function.

$$Y = AK^{\alpha}L^{\beta} \tag{6.33}$$

where

Y is the output (gross output)
L is the labor (number of workers)
K is the capital
A the other elements
α is the elasticity of capital to output
β is the elasticity of labor to output
$\alpha + \beta$ is the elasticity of productivity

The original form of the Cobb–Douglas product function reflects the unchanged returns to scale, the relationship between the final achievement and labor, the capital and other elements including natural science, knowledge of social science, the adoption of new equipment and techniques due to technology diffusion, the enhancement of labor quality, and improvement of environment and policies. We use A to represent these elements and consider it to be the technology progress or level.

Let $\alpha + \beta = 1$. From Formula 6.33 we can get

$$A = \frac{Y}{K^{\alpha}L^{\beta}} = \left(\frac{Y^{(1-\alpha)}}{L^{(1-\alpha)}}\right) \cdot \left(\frac{K^{(-\alpha)}}{Y^{(-\alpha)}}\right) = \left(\frac{Y}{L}\right)^{(1-\alpha)} \cdot \left(\frac{K}{Y}\right)^{(-\alpha)} \tag{6.34}$$

Assuming $A = A(t)$, technology progress as a function of time.

Considering the macro aspect, Formula 4.5.3 can be employed to measure the technology progress.

$$A(t) = \left(\frac{Y^{\left(1-\frac{K}{Y}\right)}}{L^{\left(1-\frac{K}{Y}\right)}}\right) \cdot \left(\frac{K^{\left(-\frac{K}{Y}\right)}}{Y^{\left(-\frac{K}{Y}\right)}}\right) = \left(\frac{Y}{L}\right)^{\left(1-\frac{K}{Y}\right)} \cdot \left(\frac{K}{Y}\right)^{\left(-\frac{K}{Y}\right)} \tag{6.35}$$

where

$A = A(t)$ is the technology progress (technology level)
Y is the gross output
L is the workers and staff members (labor)
K is the net value of fixed assets

Formula 6.35 is inducted from Formula 6.34 which can demonstrate the technology progress or technology level of the technology diffusion resource and absorber, which provides the dynamics of the technology diffusion.

6.5.3 Simulation of the Diffusion Dynamic of Technology Diffusion Field

Assume: Gross output $Y = 4{,}000{,}000$ RMB, Workers and staff members $L = 300$, net value of fixed assets $K = 5{,}000{,}000$.

The technology progress (technology level) is

$$A(t) = \left(\frac{Y}{L}\right)^{\left(1-\frac{K}{Y}\right)} \cdot \left(\frac{K}{Y}\right)^{\left(-\frac{K}{Y}\right)} = \left(\frac{4{,}000{,}000}{300}\right)^{\left(1-\frac{5{,}000{,}000}{4{,}000{,}000}\right)} \cdot \left(\frac{5{,}000{,}000}{4{,}000{,}000}\right)^{\left(-\frac{5{,}000{,}000}{4{,}000{,}000}\right)}$$

$$= 0.07041$$

Chapter 7

Technology Transfer Optimization and Disposition

Technology transfer starts from the technical achievements of the provider when he diffuses or transfers the technology to the technology adopter through various channels. Then the adopter will use, digest, absorb the new technology during the introducing process and finally make innovations. Developed countries or regions obtain great profits through technology transfer, while developing countries or regions wish to enhance their productivity level and economic strength. Hence, to fully utilize technology transfer in promoting regional economic development, every region should optimize the space available, using the power of the government and market, and make farsighted industrial development plans. In addition, based on the real situations and specific objectives of regional economic development, it is also necessary to adopt advanced technology and upgrade the industrial structure in time to promote the sustainable and healthy economic development of the country.

7.1 Linear Programming Model of Technology Transfer of Unipolar City Based on the Perfect Mechanism

It is of practical value to research the element of transfer in regional cooperation and competition within a country, to achieve optimal regional allocation and improve the efficiency of the allocation. Many domestic and foreign scholars are studying the subject extensively. However, correlative research on technology transfer through the linear programming model, on the basis of regional restrictions and conditions based on the different elements of science and technology, is lacking. There is also no reasonable explanation as to how to reach optimal allocation of spatial transfer in city network systems by means of technology transfer.

This section establishes a linear programming model of the effect of technology transfer, which treats optimal profit of technology transfer as a target, under the restrictive conditions of available resources, environmental capacity, systems constraints, and so on. It can be concluded that, with the perfect mechanism, the regional economic system which has Unipolar elements of technology and non-competitive resources, could realize the Pareto optimal distribution through profit-chasing technology transfer, whereas competitive resources cannot reach Pareto optimal distribution, and that, when the marginal income of technology transfer is equal to the marginal cost, the profit of technology transfer is maximized. The optimal allocation of technology transfer forms an echelon distribution which is restricted by profit, resources, environment, and other elements.

7.1.1 Several Important Concepts

7.1.1.1 Technology Transfer and Technology Elements Transfer

During a given time, the regional horizontal transfer of technology may result in different benefits, which essentially means the transfer and combined movement of several useful elements among different participants, under conditions of uneven distribution of resource, technology difference, etc. The relation between technology elements transfer and technology transfer is similar to that between form and effect.

Definition 7.1: (Technology transfer and technology elements transfer) Through certain approaches, local elements such as technology labor, technical information, and technology management transfer from one region to another and are rationally combined with technology elements. Technology transfer occurs when productivity has been increased. ■

Definition 7.1 has the following three meanings:

1. Technology elements carry technology transfer and are the specific manifestation of technology in the process of spatial transference. They also reflect the character of technology transfer.
2. In the process of technology transfer, technology elements of the local region combine with new elements to achieve greater productivity.
3. Technology transfer is the result of technology elements transfer.

7.1.1.2 Perfect Mechanism of Technology Elements Transfer

The mechanism of technology elements transfer differs in different economic systems. In a completely planned economic system, technology elements are utilized through planned allocation. In a complete market economy, the interest-chasing mechanism is used. History has proved that any allocation mechanism with only a single element cannot harmonize the relation between individual income and social benefit or maximize one of them. To simplify the study, we make the assumption that this kind of mechanism exists as shown in Definition 7.2.

Definition 7.2: (Perfect mechanism of technology elements transfer) The mechanism that can make reasonable allocations to individual profit and social benefit through different interests is called perfect mechanism. ■

Definition 7.2 has the following two meanings:

1. Market profit brought forth by technology elements transfer, namely individual income and social benefit, can be properly allocated by market mechanism.
2. Market mechanism cannot ensure the rational allocation of market returns (positive external effect) brought forth by technology elements transfer. In this situation, if the government can make an exact prediction of the social benefit and put forward some rational compensation mechanisms, the participants will get appropriate compensation according to their contribution to social benefit.

7.1.1.3 Competitive and Noncompetitive Technology Elements

The transfer of some technology elements, though not all, may create competition between the receiver and the provider. Also, the quantity of some technology elements being transferred may be limited. To distinguish between them, we classify them as competitive and noncompetitive elements, as shown in Definition 7.3.

Definition 7.3: (Competitive and noncompetitive technology elements) According to whether the elements transferred can harm the provider, we classify them as competitive and noncompetitive elements. ■

Definition 7.3 has the following two meanings:

1. Noncompetitive technology elements are those that, when transferred, do not harm the profit of the provider, and the quantity of which is not limited, in terms of technique, information, management, etc.
2. Competitive technology elements are those that, when transferred, can harm the provider, and the quantity of which is limited, in terms of technical talents and technical equipment.

7.1.1.4 Technology Transfer in a Unipolar City

The pursuit of profit is one of the most basic needs of people. Therefore, as long as transfer of technological elements bring profits, they will do so, and the more benefit there is, the more enthusiasm there is.

The main dynamic of technology transfer is the difference in interest between the different regions. While other conditions remain unchanged, technical elements will only move to regions with higher profit. To simplify the study, we assume that technical elements will move only to the region that provides the highest profit. In other words, we take only technology transfer of a Unipolar city into consideration, as shown in Definition 7.4.

Definition 7.4: (Technology transfer of Unipolar city) In the process of technology transfer, the central polar is the city with the highest profit. Technology transfer of a Unipolar city means that technology will be transferred only to the city with highest profit. ■

Definition 7.4 has the following two meanings:

1. Technology will be transferred from the city with a lower profit to that with a higher profit.
2. In the process of transfer, technology will be transferred to the city with the highest profit, in pursuit of maximum efficiency.

7.1.2 Linear Programming Model of Technology Transfer of a Unipolar City Based on the Perfect Mechanism

7.1.2.1 Objective Analysis

Definition 7.5: Under the condition of perfect mechanism, the objective of profit-chasing transfer of technology elements is in accordance with the optimal allocation of social benefit. ■

Proof: If technology elements are transferred with the motivation of profit-chasing, two kinds of relations may occur between individual profits and social benefit in the market economy system: individual profit may be coincident with social benefit, or it may not.

Under the first condition, the objective of profit-chasing transfer of technology elements must be in accordance with the optimal allocation of social benefit.

Under the second condition, according to Definition 7.2, perfect mechanism could make a rational compensation to the participants according to their contribution to social benefit. In this situation, the more the contribution, the more the profit they will get.

Therefore, under the conditions of perfect mechanism, the goal of profit-chasing transfer of technology elements must be in accordance with the optimal allocation of social benefit. ❑

This chapter assumes that technique-j is transferred from a Unipolar city denoted k, to m cities. The effects of technology elements transfer are maximized as an objective function, which is expressed in Formula 7.1 as follows:

$$F(\Delta P) = \max\left(\sum_i^m \sum_j^n \lambda_{k\to i,j}\,\Delta P_{k\to i,j} + \sum_i^m \sum_j^n \lambda_{i\to k,j}\,\Delta P_{k\to i,j}\right)$$
$$= \max\sum_i^m \sum_j^n (\lambda_{k\to i,j} + \lambda_{i\to k,j})\Delta P_{k\to i,j}, \quad j = 1,2,\ldots,n \text{ and } i = 1,2,\ldots,m$$

where

- $\lambda_{k\to i,j}$ is each unit benefit to city-k, from the transfer of element-j from city-k to city-i
- $\lambda_{i\leftarrow k,j}$ is each unit benefit to city-i, from the transfer of element-j from city-k to city-i
- $\Delta P_{k\to i,j}$ is the quantity of element-j from city-k to city-i
- $\lambda_{k\to i,j}\,\Delta P_{k\to i,j}$ denotes the total income of city-k, from the transfer of element-j with quantity of $\Delta P_{k\to i,j}$ from city-k to city-i
- $\lambda_{i\leftarrow k,j}\,\Delta P_{k\to i,j}$ presents the total profit of city-i, from the transfer of element-j with quantity $\Delta P_{k\to i,j}$ from city-k to city-i

7.1.2.2 Identification of Constraints

The transferring of elements is restricted by the resources available, bearing capacity of the environment, extent of system constraints, and the amount of technology elements, etc.

1. Constraints of technical condition
   a. Constraints of technical level

Through high level management, talented labor can promote the development of society with advanced tools and high technology. Thus, technology transfer happens only when the technology being introduced is of a higher level than that in the local region, and can greatly improve the production efficiency of the local region.

   b. Constraints in the number of technology elements

The number of technology elements in a certain region is limited, so the effluent technology elements cannot exceed the total number in the region.

2. Constraints of available resources

When technology elements are transferred from one region to another through certain modes and approaches, and combined with the elements already found there, production capacity is increased. After that, the process is expected to continue, although resource constraints restrict the amount of transfer.

3. Constraints of environmental bearing capacity

Technology transfer is restricted by certain environment conditions such as transportation, storage, power supply, wholesale and retail trade, finance and insurance business, scientific research services and the ability to deal with pollution. All these restrict the amount and the rate of technology transfer.

4. Constraints of the system conditions

Whether technology can be rationally transferred, effectively allocated, and play its role persistently, at least to some extent, depends on the institutional arrangements, social environment, and cultural atmosphere, which have to be conducive to developing innovative activities and finding one's potential talent. Furthermore, the requirements of different types of technology are not the same. Systems conditions generally include tax policy, distribution and incentive mechanisms, security assurances, and interest returns, as well as some other aspects.

5. Constraints of profit conditions

In perfect mechanism, "input–output benefit" is the trend for the transfer of technology elements. Whether the resources are competitive or noncompetitive, technology elements will be transferred only when the profits got from transferring are greater than before.

According to the above analysis, we assume that they are linearly related to simplify the study, and the optimization model of technology transfer is established as follows:

$$F(\Delta P) = \max\left(\sum_i^m \sum_j^n \lambda_{k\to i,j}\,\Delta P_{k\to i,j} + \sum_i^m \sum_j^n \lambda_{i\to k,j}\,\Delta P_{k\to i,j}\right)$$
$$= \max \sum_i^m \sum_j^n (\lambda_{k\to i,j} + \lambda_{i\to k,j})\Delta P_{k\to i,j} \tag{7.1}$$

$$
\text{s.t.}\quad
\begin{cases}
r_{i,j}(t) = a_{i,j}(t)P'_{i,j} \le R_{i,j}(t) \\
e_{i,j}(g) = b_{i,j}(g)P'_{i,j} \le E_{i,j}(g) \\
s_{i,j}(c) = h(\Delta P_{k\to i,j}) \le S_{i,j}(c) \\
P'_{i,j} = P_{i,j} + \Delta P_{k\to i,j} \\
\sum_{i}^{m} \Delta P_{k\to i,j} \le P_{k,j} \\
l(P_{k,j}) = l(\Delta P_{k\to i,j}) > l(P_{i,j}) \\
\lambda_{k\to i,j} + \lambda_{i\to k,j} > 0 \\
r_{i,j}(t), e_{i,j}(g), s_{i,j}(c), \Delta P_{k\to i,j} \ge 0
\end{cases}
\tag{7.2}
$$

where

$P'_{i,j}$ is the quantity of element-j after transfer of element has accomplished

$P_{k,j}$ is the quantity of element-j before transfer of element starts

$r_{i,j}(t)$ is the amount of resources needed in the process of using the element-j in city-i

t denotes the category of element

$a_{i,j}(t)$ is the amount of resources needed for each unit element

$R_{i,j}(t)$ is the amount of resources that city-i could provide for element-j

$e_{i,j}(g)$ is the demand of production environment in the process of using the element-j in city-i

g denotes the category of element

$b_{i,j}(g)$ is the amount of resources needed for each unit element

$E_{i,j}(g)$ is the capability that city-i could bear for element-j

$s_{i,j}(c)$ is the system guarantee demanded in the process of using the element-j in city-i, and it is the function of $\Delta P_{k\to i,j}$, here c denotes the category of system restraints

$S_{i,j}(c)$ is the total system restrictions of city-i that is relational with local investment policy and the policy of element transfer and so on

$l(P_{k,j})$ is the level of scientific and technological element-j of city-k

$l(P_{i,j})$ is that of city-k. They are the function of $\Delta P_{k\to i,j}$

7.1.3 Study of the Properties of the Linear Programming Model of Technology Transfer of a Unipolar City Based on Perfect Mechanism

To conduct research on the properties of technology transfer in a Unipolar city on the basis of perfect mechanism, we put forward the following two lemmas.

Lemma 7.1: The receiver region can realize the maximum profit from elements transfer. ■

Proof: The profit of receiver city-i from elements transfer is, $F(i)=\lambda_{i\leftarrow k}\Delta P_{k\to i}=\lambda_i X_i$, the restrictive condition is $D(i)=\{X_i\,|\sum_{j=1}^{n} p_j x_j = d, x_j \geq 0\}$. Then, we get optimal profit.

1° Certify that the condition of the linear programming model is a convex set.

To prove that the set of all feasible solutions is a convex set, it is enough to prove that the point in the line which joins random points belong to $D(i)$.

Assume $X^{(1)}=(x_1^{(1)},x_2^{(1)},\ldots,x_n^{(1)})^{\mathrm{T}}$, $X^{(2)}=(x_1^{(2)},x_2^{(2)},\ldots,x_n^{(2)})^{\mathrm{T}}$ are random points belonging to $D(i)$, and $X^{(1)}\neq X^{(2)}$.

So

$$\sum_{j=1}^{n} p_j x_j^{(1)} = b, \quad x_j^{(1)} \geq 0, \quad j=1,2,\ldots,n$$

$$\sum_{j=1}^{n} p_j x_j^{(2)} = b, \quad x_j^{(2)} \geq 0, \quad j=1,2,\ldots,n$$

Let $X=(x_1,x_2,\ldots,x_n)^{\mathrm{T}}$ be the point in the line which joins $X^{(1)}$ and $X^{(2)}$, and then $x_j=\alpha x_j^{(1)}+(1-\alpha)x_j^{(2)}$, $(0\leq\alpha\leq 1)$ are the constraints, so that

$$\begin{aligned}\sum_{j=1}^{n} p_j x_j &= \sum_{j=1}^{n} p_j[\alpha x_j^{(1)}+(1-\alpha)x_j^{(2)}] \\ &= \alpha\sum_{j=1}^{n} p_j x_j^{(1)} + \sum_{j=1}^{n} p_j x_j^{(2)} - \alpha\sum_{j=1}^{n} p_j x_j^{(2)} \\ &= \alpha d + d - \alpha d = d\end{aligned}$$

And because $x_j^{(1)}, x_j^{(2)} \geq 0$, $\alpha > 0$, $1-\alpha > 0$, so, $x_j \geq 0$, $i=1,2,\ldots,n$ $X\in D(i)$, $D(i)$ is a convex set.

2° Due to the limitation of resources and bearing capability of the environment, the region where linear programming is feasible is limited.

According to the theorem of linear programming, if the feasible region is a limited convex set, the objective function must reach the maximum at the points of the feasible region. Hence, the maximum profit of the receiver city-i from elements transfer can be obtained. ❑

Lemma 7.2: The profit of the transmitter region from elements transfer can reach the maximum. ■

Proof: The profit of transmitter city-k from elements transfer is, $F(k) = \lambda_{k\to i}\Delta P_{k\to i} = \lambda_k X_k$, the restrictive condition is $D(k) = \{X_k = \sum_{i=1}^{m} X_i,\ X_i \geq 0\}$. Then city-$k$ could gain maximum or minimum profit.

Let $\lambda_i = 1$ in 1° of Lemma 7.1, then X_i can reach the maximum at the points of the feasible region. So X_k can reach the maximum if $D(k) = \{X_k = \sum_{i=1}^{m} X_i,\ X_i \geq 0\}$. So, if $\lambda_k \geq 0$, $F(k)$ can reach the optimal point when all receiver regions gain the most profit. When $\lambda_k < 0$, $F(k)$ is minimized. ❑

Proposition 7.2: In an economic system with a unipolar region of technology based on perfect mechanism, noncompetitive resources can realize Pareto optimal distribution by technology elements profit-chasing transfer, whereas competitive resources cannot realize Pareto optimal distribution. ■

The proposition shows two situations.

In the first situation, noncompetitive resources can realize Pareto optimal distribution by technology elements profit-chasing transfer.

Based on perfect mechanism, the transfer of noncompetitive elements does not affect the profit of the transmitter region, $\lambda_{k\to i} \geq 0$, $\lambda_{i\leftarrow k} \geq 0$, so by Lemma 7.1 and Lemma 7.2, when the receiver region achieves the most profit, the transmitter region can also gain the optimal profit. Then the profit of elements allocation will reach the maximum, and Pareto optimal distribution will be realized. ❑

In the second situation, competitive resources cannot realize Pareto optimal distribution. This can be proved by reduction to absurdity.

Assume that competitive resources can reach Pareto optimal distribution by technology elements profit-chasing transfer. In other words, the individual in society gets maximum profit by not damaging the interest of others.

In perfect mechanism, the transfer of competitive resources can cut down the profit of the transmitter, that is, $\lambda_{i\leftarrow k} \geq 0$, $\lambda_{k\to i} \leq 0$, so an increase in $F(i)$ will make $F(k)$ go down. That is, the increase in the profit of the receiver region can make that of the transmitter region decrease, so this is not a Pareto optimal state, and so it is impossible to get Pareto optimal distribution. ❑

Proposition 7.3: On the condition of perfect mechanism, competitive or noncompetitive resources will improve the total profit. ■

Proof: It could be proved by reduction to absurdity.

Under conditions of perfect mechanism, competitive or noncompetitive resources will not increase the total profit through transfer.

The profit of competitive and noncompetitive resources from transferring is

$$\max\left(\sum_i^m\sum_j^n \lambda_{k\to i,j}\,\Delta P_{k\to i,j} + \sum_i^m\sum_j^n \lambda_{i\to k,j}\,\Delta P_{k\to i,j}\right)$$
$$= \max\sum_i^m\sum_j^n (\lambda_{k\to i,j} + \lambda_{i\to k,j})\Delta P_{k\to i,j} < 0$$

$$\text{And } (\lambda_{k\to i,j} + \lambda_{i\to k,j}) \geq 0, \quad \Delta P_{k\to i,j} \geq 0$$

$$\text{So } \max\sum_i^m\sum_j^n (\lambda_{k\to i,j} + \lambda_{i\to k,j})\Delta P_{k\to i,j} \geq 0$$

which is not in accordance with the assumption, so the assumption is wrong, and the proposition is proved. ❑

Proposition 7.4: The optimal allocation of technology transfer constitutes an echelon distribution which is restricted by profit, resource, environment, and other elements. ■

From Proposition 7.2, we know that for region i, the optimal profit is $\{F(i)\}$ under the condition of $D_i = \left\{X_i \middle| \sum_{j=1}^n p_j x_j = b, x_j > 0\right\}$ by transferring its technology X_i.

Then, the optimal profit for all of regions is $\max(F) = \{F_1, F_2 \cdots F_i \cdots F_m\}$, under the condition of $D = \{D_1, D_2 \cdots D_i \cdots D_m\}$, by transferring their technologies $X = \{X_1, X_2 \cdots X_i \cdots X_m\}$.

Therefore, the optimal allocation of technology transfer constitutes an echelon distribution which is restricted by profit, resource, environment, and other elements.

When restrictive conditions are the same for all the regions, namely $D_1 = D_i$, $i = 2, 3, \ldots, m$, and the number of elements of the different regions is $X_1 = X_i$, $i = 2, 3, \ldots, m$, there is uniform distribution of technology transfer. The proposition is proved. ❑

Proposition 7.5: Under the condition of perfect mechanism, the effect from technology elements transfer can reach the maximum when the marginal profit of elements transfer equals marginal cost. ■

Proof: It is assumed that the profit of technology elements is $Ip_{k\to i,j} = v(\Delta P_{k\to i,j})$, and the cost is $Ce_{k\to i,j} = w(\Delta P_{k\to i,j})$. Then the effect is

$$\pi(\Delta P_{k\to i,j}) = Ip_{k\to i,j} - Ce_{k\to i,j}$$
$$= v(\Delta P_{k\to i,j}) - w(\Delta P_{k\to i,j})$$

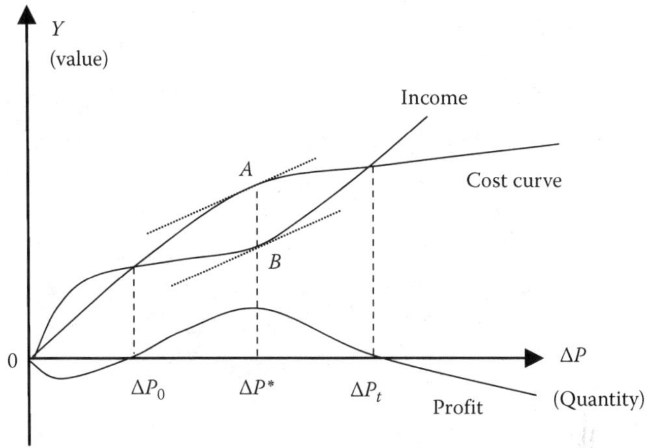

Figure 7.1 Relationship of marginal income, marginal cost, and quantity of flow.

With universality, suppose the cost and the income curves are shown in Figure 7.1. The slopes represent the marginal income and the marginal cost, respectively.

When $\pi(\Delta P_{k\to i,j})>0$, the transfer of technology elements occurs. From Figure 7.1, it is obvious that technology elements are transferred at point ΔP_0, the maximum profit of transfer is reached at point ΔP^*, and the transfer of elements stops at point ΔP_t.

For finding the maximum value of $\pi(\Delta P_{k\to i,j})$, let $\pi'(\Delta P_{k\to i,j}) = 0$, that is, $\frac{\Delta \pi}{\Delta P} = \frac{\Delta v}{\Delta P} - \frac{\Delta w}{\Delta P} = 0$, and we get the solution ΔP^*. So the maximum profit is realized when marginal income is equal to marginal cost. ❑

7.1.4 Case Study

Assume that A, B, C, D, E, F, G, are the server cities, of which city-A possesses of the maximum and the highest level of a certain element of science and technology. This element is of much value in the development of regional economies, and it is sought by every city. So during the transfer of this element, a positive system is provided, and so the constraint of system is not found in the model. According to the model discussed above, the transfer of this element is described here.

7.1.4.1 Linear Programming Model for Noncompetitive Resource

Assume the parameters that are provided in Tables 7.1 and 7.2.

Using the optimization model, the transfer and the profit of every city can be worked out, as shown in Figure 7.2.

Table 7.1 Indicators of Element Transfer of Every City

	A	B	C	D	E	F	G
$\lambda_{A\to i}$		2.8	2.2	2.5	1.7	1.4	1.1
$\lambda_{i\to A}$		4.5	4	3.5	3	2.5	2
P_i		93	80	85	90	73	65
$\Delta P_{A\to i,j}$		7	0	6.67	5.83	15.46	25
$\lambda_{A\to i,j}\Delta P_{A\to i,j}$	95.33	19.6	0	16.68	9.91	21.64	27.5
$\lambda_{i\to A,j}\,\Delta P_{A\to i,j}$		31.5	0	23.35	17.49	38.65	50

Table 7.2 Indicators of Element Transfer of Every City

	$a_i(t)$			$R_i(t)$			$b_i(t)$			$E_i(m)$		
	$a(1)$	$a(2)$	$a(3)$	$R(1)$	$R(2)$	$R(3)$	$b(1)$	$b(2)$	$b(3)$	$E(1)$	$E(2)$	$E(3)$
B	0.12	0.15	0.2	15	18	25	0.24	0.26	0.3	25	28	30
C				18	15	16				24	27	30
D				18	14	25				22	24	35
E				20	16	20				23	26	32
F				15	14	18				25	23	28
G				12	15	20				22	25	27

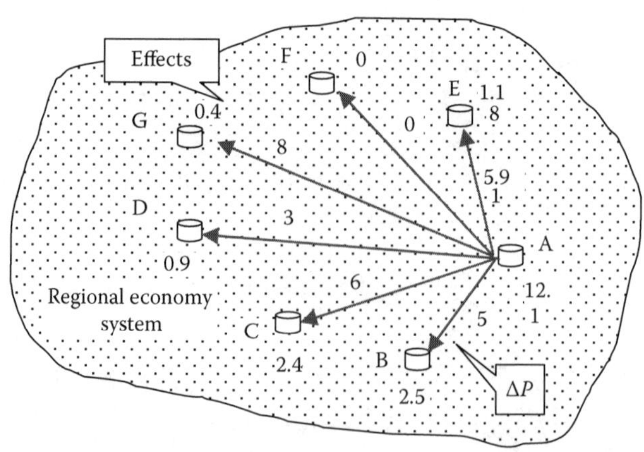

Figure 7.2 Profit and transfer of technical elements in every city.

Table 7.3 Indicators of Element Transfer of Every City

	A	*B*	*C*	*D*	*E*	*F*	*G*
$\lambda_{A\to i}$		−2.8	−2.2	−2.5	−1.7	−1.4	−1.1
$\Delta P_{A\to i,j}$		7	0	6.67	5.83	15.46	5.04
P_i	40	93	80	85	90	73	65
$\lambda_{A\to i,j}\ \Delta P_{A\to i,j}$	−73.37	−19.6	0	−16.68	−9.91	−21.64	−5.54
$\lambda_{i\to A,j}\ \Delta P_{A\to i,j}$	121.07	31.5	0	23.35	17.49	38.65	10.08

7.1.4.2 Linear Programming Model for Competitive Resource

Assume the parameters are those provided in Table 7.3, and that the other coefficients are the same as above.

7.2 Linear Programming Model of Technology Transfer of Unipolar City Based on the Perfect Mechanism

We have talked about the linear programming model of technology transfer under unipolar condition. Actually, technology transfer has several objectives and a multipolar feature; therefore this part will focus on the linear programming model of technology transfer of a multipolar city based on the perfect mechanism.

Definition 7.6: (Network transfer of urban technology) In the process of urban technology transfer, the elements of technology will transfer to cities with higher profit. These subjects, either the supplier or the demand party, constitute a dynamic network structure, details as shown in "Mechanism of technology transfer." ■

7.2.1 Linear Programming Model of Technology Transfer of Multipolar City Based on the Perfect Mechanism

7.2.1.1 Objective Analysis

Technology transfer is intended to better realize the value of the technology elements as well as to ensure the prosperity, stability, and harmonious development of society. On this basis, the objectives during the technology transfer between *n* cities are assumed to be as follows:

1. Objective one: Profit maximization of the transfer of the technology elements

Pursuing profit is people's basic demand. The transfer of the technology elements takes place once benefit is assured during the process and profit is increased by the transfer dynamic. Here the transfer profit indicates the net profit, subtracting the profit of the transmitter region and the transfer cost from the profit of the receiver region.

$$F(\Delta P_1) = \max\sum_{i=1}^{n}\left(\frac{1}{2}\sum_{k=1}^{n}\lambda_{k\to i}\Delta P_{k\to i}\right)$$

$$= \frac{1}{2}\max\sum_{i=1}^{n}\sum_{k=1}^{n}(\lambda_{k\to i}\Delta P_{k\to i}) \quad (7.3)$$

where $\lambda_{k\to i}$ is each unit profit to city-k, from the transfer of elements from city-k to city-i.

This is the difference between the profit and cost for the transfer of each unit of technology element. When it transfers from city-k to city-i, $\lambda_{k\to i}$ is positive, from city-i to city-k, $\lambda_{k\to i}$ is negative.

$\Delta P_{k\to i}$ is the quantity of element from city-k to city-i. When it transfers from city-k to city-i, $\Delta P_{k\to i}$ is positive, in the reverse order, $\Delta P_{k\to i}$ is negative, and $\Delta P_{i\to i} = 0$.

$\lambda_{k\to i}\Delta P_{k\to i}$ denotes the total profit of city-k, from the transfer of element with quantity of $\Delta P_{k\to i,j}$ from city-k to city-i.

2. Objective two: The coordinated and steady development in the region

The coordinated and steady development in the region is the important goal for all governments when developing regional economy. It reflects the reasonable and equal distribution of various political and economic interests among all social members. As an inevitable result of market economy, this objective demands technology transfer. Here it is assumed that the sum of the technology growth rate of all regions is largest.

$$F(\Delta P_2) = \max\sum_{i=1}^{n} t_i = \max\sum_{i=1}^{n}\frac{\sum_{k=1}^{n}\Delta P_{k\to i}}{P_i} \quad (7.4)$$

where P_i is the original quantity of technology element in region-i, $t_i = \frac{\sum_{k=1}^{n}\Delta P_{k\to i}}{P_i}$.

3. Objective three: Regional stability

Maintaining the stability of the region is an important goal of the development plan and an important content of the comprehensive construction of a well-off society in China. Let $t_i = \frac{\sum_{k=1}^{n} \Delta P_{k \to i}}{P_i}$ be the technology growth rate in different regions, then variance of t_i and the average rate $\bar{t}$ is a minimum.

$$F(\Delta P_3) = \min \sum_{i=1}^{n} (t_i - \bar{t})^2 \tag{7.5}$$

where $\bar{t}$ is the average growth rate in different regions, $\bar{t} = \frac{\sum_{i=1}^{n} t_i}{n}$.

7.2.1.2 Constraint Conditions

The quantity of technology transfer is restricted by available resources, environment bearing capacity, degree of constraint in the system, quantity of elements, technical level, and other conditions. For convenience, it is assumed that these are linear relationships.

According to the analysis above, the optimal model of technology transfer is as follows:

$$F(\Delta P_{k \to i}) = \max \left\{ \frac{1}{2} \alpha_1 \sum_{i=1}^{n} \sum_{k=1}^{n} (\lambda_{k \to i} \Delta P_{k \to i}) + \alpha_2 \sum_{i=1}^{n} t_i - \alpha_3 \sum_{i=1}^{n} (t_i - \bar{t})^2 \right\}$$

$$\text{s.t.} \begin{cases} P_i + \sum_{i=1}^{n} \Delta P_{k \to i} \geq 0 \\ r_i(t) = a_i(t) P_i' \leq R_i(t) \\ e_i(g) = b_i(g) P_i' \leq E_i(g) \\ s_i(c) = h(\Delta P_{k \to i}) \leq S_i(c) \\ l(P_k) = l(\Delta P_{k \to i}) > l(P_i) \\ t_i = \dfrac{\sum_{k=1}^{n} \Delta P_{k \to i}}{P_i} \\ \alpha_1 + \alpha_2 + \alpha_3 = 1 \\ r_{i,j}(t), e_{i,j}(g), s_{i,j}(c), \alpha_1, \alpha_2, \alpha_3 \geq 0 \end{cases} \tag{7.6}$$

where

$P_i' = P_i + \sum_{k=1}^{n} \Delta P_{k \to i}$ is the present quantity of elements in region-i

$r_i(t)$ is the other resources necessary for the formation of technology in region-i

t represents the sort of resources required

$a_i(t)$ is the consumption coefficient
$R_i(t)$ is the available quantity of resources relating to local resource environment
$e_i(g)$ is the product environment required for the formation of technology in region-i
g is the indicator of the environment bearing capacity
$b_i(g)$ indicates the bearing coefficient
$E_i(g)$ is the maximum bearing capacity of the environment, which is related to the local resource environment
$s_i(c)$ is the system guarantee required in region-i and is relational with the nature of the transfer technology
c is the binding clause of systems
$S_i(c)$ is the constraint of systems related to the local policies of attracting investment and technology transfer
$l(P_k)$ denotes the technical level in region-k
$l(P_i)$ represents the level of the technology element in region-i

7.2.2 Study of the Properties of the Model of Multiobjective Programming of Technology Transfer Based on the Perfect Mechanism

To study the technology transfer based on the perfect mechanism, the following propositions are advanced:

Proposition 7.6: Without considering the stability of regions, meaning $\alpha_3 = 0$, the technology transfer will bring about improvement in the integral profit and regional coordinated development. Also, the optimal allocation of technology transfer constitutes an echelon distribution which is restricted by profit, resource, environment, and some other elements. ■

Proof: (1) Without considering the stability of the regions, meaning $\alpha_3 = 0$, the technology transfer will bring about improvement in the integral profit and regional coordinated development.

In the objective function, when $\alpha_3 = 0$, it turns into a single-objective linear program of $\Delta P_{k\to i}$.

$$F(\Delta P_{k\to i}) = \max\left\{\frac{1}{2}\alpha_1 \sum_{i=1}^{n}\sum_{k=1}^{n}(\lambda_{k\to i}\Delta P_{k\to i}) + \alpha_2 \sum_{i=1}^{n}\frac{\sum_{k=1}^{n}\Delta P_{k\to i}}{P_i}\right\}$$

The objective of the function is $F(\Delta P_{k\to i}) = \beta_{i\to k}\,\Delta P_{k\to i} = \lambda_i X_i$; the constraint condition is $D(i) = \left\{X_i \left| \sum_{j=1}^{n} p_j x_j d, x_j \geq 0\right.\right\}$.

1. To certify that the condition of the linear programming model gives rise to a convex set.

To prove that the set of all feasible solutions is a convex set, it is enough to prove that the point in the line which joins random points belong to $D(i)$.

Assuming $X^{(1)} = (x_1^{(1)}, x_2^{(1)}, \ldots, x_n^{(1)})^T$, $X^{(2)} = (x_1^{(2)}, x_2^{(2)}, \ldots, x_n^{(2)})^T$ are random points belonging to $D(i)$, and $X^{(1)} \neq X^{(2)}$,

So

$$\sum_{j=1}^{n} p_j x_j^{(1)} = b, \quad x_j^{(1)} \geq 0, \quad j = 1,2,\ldots,n$$

$$\sum_{j=1}^{n} p_j x_j^{(2)} = b, \quad x_j^{(2)} \geq 0, \quad j = 1,2,\ldots,n$$

Let $X = (x_1, x_2, \ldots, x_n)^T$ be the point in the line which joins $X^{(1)}$ and $X^{(2)}$, and $x_j = \alpha x_j^{(1)} + (1-\alpha)x_j^{(2)}$, $(0 \leq \alpha \leq 1)$ be the constraint, so that

$$\sum_{j=1}^{n} p_j x_j = \sum_{j=1}^{n} p_j[\alpha x_j^{(1)} + (1-\alpha)x_j^{(2)}]$$

$$= \alpha \sum_{j=1}^{n} p_j x_j^{(1)} + \sum_{j=1}^{n} p_j x_j^{(2)} - \alpha \sum_{j=1}^{n} p_j x_j^{(2)}$$

$$= \alpha d + d - \alpha d = d$$

And therefore $x_j^{(1)}, x_j^{(2)} \geq 0, \alpha > 0, 1-\alpha > 0$, so, $x_j \geq 0$, $i = 1,2, \ldots, n$. Hence $X \in D(i)$, $D(i)$ is convex set.

2. Due to the limitation of resources and bearing capability of environment, the feasible region of the linear program is limited.

By the theorem of the linear programming, if the feasible region is a limited convex set, the objective function must reach maximum at the points of the feasible region, so that, the maximum effect of receiver city-i from elements transfer could be realized.

In conclusion, the profits of technology transfer in every region could reach the maximum value; therefore the transfer would undoubtedly improve the integral profits during the process of technology transfer.

Proof: (2) Without considering the stability of regions, meaning $\alpha_3 = 0$, the optimal allocation of technology transfer constitutes an echelon distribution which is restricted by profit, resource, environment, and some other elements.

From Proof (1), we know that the element of technology transfer X_i will reach the maximum profit, max $\{F(i)\}$ under the condition of $D_i = \left\{X_i \middle| \sum_{j=1}^{n} p_j x_j = b, x_j > 0\right\}$.

Therefore, the elements $X = \{X_1, X_2 \cdots X_i \cdots X_m\}$ will reach max $(F) = \{F_1, F_2 \cdots F_i \cdots F_m\}$ on the condition of $D = \{D_1, D_2 \cdots D_i \cdots D_m\}$.

So we can prove that the optimal allocation of technology transfer constitutes an echelon distribution which is restricted by profit, resource, environment, and some other elements.

When the constraint conditions of the technology transfer in all cities remain the same, that is $D_1 = D_i$, $i = 2, 3, \ldots, m$, then the quantity $X_1 = X_i$, $i = 2, 3, \ldots, m$ in these cities will constitute an even distribution of the element of technology transfer. It is a special case of this proposition. ❑

Proposition 7.7: Considering only the regional stability, that is when $\alpha_3 = 1$ and $\alpha_1 = 0$, $\alpha_2 = 0$, then it will achieve the optimal effects of regional stability.

$$F(\Delta P_3) = \min \sum_{i=1}^{n} (t_i - \bar{t})^2$$

■

Proof: When $\alpha_1 = 0$, $\alpha_2 = 0$, $\alpha_3 = 1$ in the objective function, this function becomes the objective programming of coordinated regional development.

Proof: Hessian matrix of the known objective function is

$$H(X^*) = \begin{bmatrix} \dfrac{\partial^2 F(X^*)}{\partial x_1^2}, \dfrac{\partial^2 F(X^*)}{\partial x_1 \partial x_2}, \ldots, \dfrac{\partial^2 F(X^*)}{\partial x_1 \partial x_n} \\ \dfrac{\partial^2 F(X^*)}{\partial x_2 \partial x_1}, \dfrac{\partial^2 F(X^*)}{\partial x_2^2}, \ldots, \dfrac{\partial^2 F(X^*)}{\partial x_2 \partial x_n} \\ \cdots \\ \dfrac{\partial^2 F(X^*)}{\partial x_n \partial x_1}, \dfrac{\partial^2 F(X^*)}{\partial x_n \partial x_2}, \ldots, \dfrac{\partial^2 F(X^*)}{\partial x_n^2} \end{bmatrix}$$

Here $\dfrac{\partial^2 F(X^*)}{\partial t_i^2} = 2(i = 1, 2, \ldots, n)$, $\dfrac{\partial^2 F(X^*)}{\partial t^2} = 2n$

then

$$\frac{\partial^2 f(X^*)}{\partial P_i'^2} \geq 0, \quad i = 1, 2, \ldots, n, \quad \frac{\partial^2 F(X^*)}{\partial t^2} = 2n > 0$$

For $\dfrac{\partial^2 f(X^*)}{\partial t_i \partial t} = -2, \quad i = 1, 2, \ldots, n$.

Therefore, the Hessian matrix is

$$H(X^*) = \begin{bmatrix} 2,-2,\ldots,-2 \\ -2,2,\ldots,-2 \\ \ldots \\ -2,-2,\ldots,2n \end{bmatrix}$$

Thus, the Hessian matrix at the point of X^* is positive semidefinite and it can be concluded that the objective function is a convex function. Furthermore, the optimal model of regional stability is a convex model. Due to the limitation of resources and the bearing capability of the environment, the feasible region of linear programming is limited. By the theorem of linear programming, if the feasible region is a limited convex set, the objective function must reach maximum at the points within the feasible region, ensuring that the optimal effect of regional stability is realized. ❑

Proposition 7.8: When $k = 1$, $\Delta P_{k \to i} \geq 0$, $\alpha_3 = 1$, $\alpha_1 = 0$, $\alpha_2 = 0$ the model changes into unipolar-based technology transfer on perfect mechanism and has the features of this technology transfer. ■

Proof: When $k = 1$, $\Delta P_{k \to i} \geq 0$, there is only one source of technology, that is unipolar: $\alpha_3 = 1$, $\alpha_1 = 0$, $\alpha_2 = 0$. And the model has a single objective. Therefore the model changes into unipolar technology transfer based on perfect mechanism.

Unipolar technology transfer based on perfect mechanism has the following features:

1. During technology transfer, the profit of the receiver region can reach the maximum value.
2. During technology transfer, the transferring benefit of the transmitter region can reach the extreme value.
3. On the perfect mechanism, in the regional economic system with unipolar elements of technology, noncompetitive resources could realize the Pareto optimal distribution by profit-making technology elements transfer, whereas competitive resources cannot reach Pareto optimal distribution.
4. On the perfect mechanism, the transfer will improve integral benefits, irrespective of whether the resources are competitive or noncompetitive.
5. The optimal allocation of technology transfer constitutes an echelon distribution which is restricted by profit, resource, environment, and some other elements. ❑

7.2.3 Case Study

7.2.3.1 Interpretation of Data

It is assumed that there are seven cities, A, B, C, D, E, F, and G. The transfer of a technology element among them according to the model discussed earlier is the subject of this study. The indicator values of the linear programming model of technology transfer are shown in Tables 7.4 through 7.6.

7.2.3.2 Calculation Results

The technology transfer and the profits among the five cities calculated based on the above model are shown in Figure 7.3.

Table 7.4 Indicators of One Element of Technology Transfer among Cities (Year: 2006)

City	A	B	C	D	E
Price of a technology element	0.9	1.05	1.48	1.9	0.67
$\lambda_{A\to i}$		0.1	0.4	0.8	−0.2
$\lambda_{i\to A}$		−0.05	−0.37	−0.66	0.15
$\lambda_{B\to i}$			0.4	0.7	−0.3
$\lambda_{i\to B}$			−0.3	−0.6	0.28
$\lambda_{C\to i}$				0.4	−0.8
$\lambda_{i\to C}$				−0.3	0.7
$\lambda_{D\to i}$					−1.1
$\lambda_{i\to D}$					1
P_i	12,775,000	3,698,910,000	705,545,000	0	702,210,000
$\Delta P_{A\to i,j} + \Delta P_{i\to A,j}$	0	−9,753,000	−9,766,000	2,451,000	−44,642,000
$\Delta P_{B\to i,j} + \Delta P_{i\to B,j}$	9,753,000	0	7,314,000	74,656,000	−2,207,650
$\Delta P_{C\to i,j} + \Delta P_{i\to C,j}$	9,766,000	−7,314,000	0	1,064,200	−2,438,000
$\Delta P_{D\to i,j} + \Delta P_{i\to D,j}$	−2,451,000	−74,656,000	−1,064,200	0	−12,387,630
$\Delta P_{E\to i,j} + \Delta P_{i\to E,j}$	44,642,000	2,207,650	2,438,000	12,387,630	0

Table 7.5 Minimum Resource of One Technology Element Required in Different Cities

City	*A*	*B*	*C*	*D*	*E*
Quantity of a technology element	59,100,000	2,175,670,000	417,030,000	97,630,000	642,035,000
Quantity of a technology element (bearing capacity of the environment)	65,010,000	3,698,910,000	705,545,000	107,393,000	702,210,000

Table 7.6 Indicators of One Element of Technology Transfer among Cities

	$a_i(t)$			$R_i(t)$		
	a(1)	*a(2)*	*a(3)*	*R(1)*	*R(2)*	*R(3)*
B	2.379×10^{-9}	5.289×10^{-5}	8.8×10^{-6}	11.36	5294.69	18,892.463
C				110.92	3769.355	21,995.888
D				10.06	3889.805	18,940.436
E				125.8	3029.50	12,852.913
	$b_i(t)$			$E_i(m)$		
	b(1)	*b(2)*	*b(3)*	*E(1)*	*E(2)*	*E(3)*
	2.312×10^{-9}	3.456×10^{-5}	4.351×10^{-6}	129.95	60,888.935	2,172,633.245
				1275.58	43,347.5825	252,952,620
				122.59	44,729.3075	2,178,150.14
				1446.7	34,839.25	1,478,084.65

Notes: (1) *R*(1), *R*(2), and *R*(3), represents the restriction of three resources respectively on one technology element. (2) *a*(1), *a*(2), and *a*(3), represents the consumption coefficient of three resources respectively that one technology element demands.

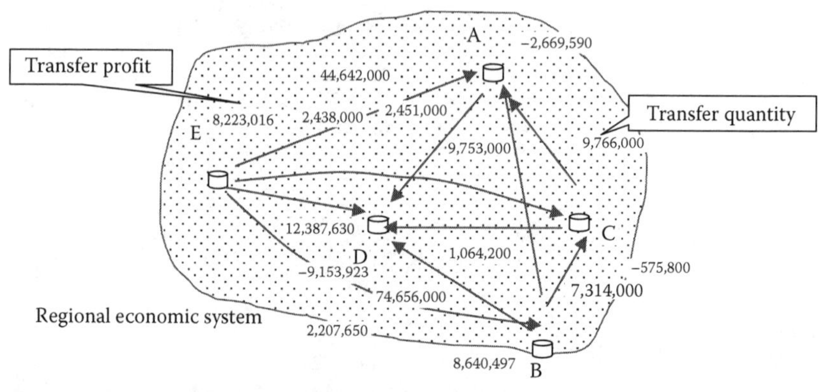

Figure 7.3 Diagram illustrating technology transfer and the profits among cities.

7.2.3.3 Analysis of the Results

1. Under the conditions of perfect mechanism, "input–output benefit" becomes the guide to technology transfer. Competitive technology elements first move to the region that will bring it the most profit and then move to other regions along with the value of the transfer. As a result of this, polarization comes into being.
2. The restrictive conditions of the receiver region would constrain the quantity of element transfer, and lack of some kinds of resources can becomes the bottleneck of the transfer. Therefore, a solution to the problem of the bottleneck is the first task.
3. Under the same factors of influence, technology elements are intended for regions with higher output benefits, and the quantity of transfer increases with the difference of interest. Therefore, some less developed regions in the West must ensure enough comparable benefits to enhance the attraction for technology elements.
4. In unipolar technology transfer, competitive resources first transfer to the region with maximum "transmitting profits." When they reach the maximum transmitting capacity restricted by this region, they will turn to other regions successively according to the value of their transmitting incomes. The transmitting quantity of noncompetitive resources is totally under the constraint of the resources, environment, systems, and other elements of the receiver city.
5. In unipolar technology transfer the transfer of technology elements in pursuit of benefits forms an echelon of distribution of productivity transfer, and changes the space structure of productivity to a larger extent.

7.3 Measurement Model of Government Matching Funds Based on Maximum Social Benefits

Technology transfer achieves its objective when it translates into goods produced. Introducing technology at a reasonable point in time has the advantage of reducing risks and costs in R&D, improving productivity, accessing technology benefit monopoly, playing a spillover effect, incorporating other enterprises' imitation and development, and enhancing the region's core competitiveness.

The government plays an important role in the process of technology introduction. It is not only the decision-maker in technology introduction, but also the beneficiary of spillover benefits from the introduction of technology. How the government should help in the introduction of technology has been a matter of concern for a large number of scholars and become a popular topic in research. Based on the endogenous economic theory, this book discusses the technology output function in the introduction of technology and suggests the optimization model with matching funds from the government. Under the restricted conditions of available resources, investment funds, and institutional constraints, the book researches how the government collocates its technology introduction funds rationally to achieve the best possible results.

7.3.1 Government Investment and Planning Model in Technology Introduction

7.3.1.1 Regional Production Function Model

Technology is a kind of intellectual property and a process of input–output, where the cost of technology introduction is input and the improvement in production benefit is output. The business and government need to pay a certain cost when introducing technology. To increase productivity, seize advantage of technology, and achieve technical effective monopoly, enterprises will actively introduce advanced technology. At the same time the introduced technology will have an effect on the main body of transfer, and also have a spillover effect on other social bodies such as the imitation and cooperation in technology among enterprises. To obtain the technology spillover benefits and enhance regional comprehensive competitiveness, the government will pay the cost to import advanced technology. The definitions of the payment cost of enterprises' introduced technology, the payment cost of governments' introduced technology and the social benefits of introduced technology are given as Definitions 7.7 through 7.9.

Definition 7.7: (Payment cost of enterprises' introduced technology) The payment cost of enterprises' introduced technology refers to the technology price paid by enterprises to the suppliers and is based on the expected profit by introducing certain techniques. ■

Definition 7.8: (Social benefits of introduced technology) The social benefits of introduced technology refers to the sum of benefits in business and society after introducing technology. ■

Definition 7.9: (Payment cost of governments' introduced technology) The payment cost of governments' introduced technology refers to the technology price paid by the government to the suppliers, based on the expected external spillover benefits by introducing certain techniques in a region which does not include the benefits of enterprises' investment. ■

With endogenous growth theory, its output is the function of the investments of every industry in the region.

1. When industry i does not introduce technology in a certain region, the production function is shown in Formula 7.7.

$$Y_i = A_i e^{\sigma_i t} \cdot L_i^{\alpha_i} \cdot K_i^{\beta_i} \tag{7.7}$$

where

Y represents the added value of the industry
L and K represent the labor and capital input, respectively
$Ae^{\sigma t}$ represents the Solow remainder term
σ represents the natural growth rate of technology in this industry
α represents labor output elastic coefficient of this industry
β represents capital output elastic coefficient of this industry
i represents the ith industry in a certain region

2. At the same time, the industry i introduces this technology in the region, and the production function is shown in Formula 7.8.

$$Y_i' = A_i e^{\sigma_i t} \cdot L_i^{\alpha_i} \cdot K_i^{\beta_i} \cdot (P_{Ei} + P_{Gi})^{\lambda_i} \tag{7.8}$$

where

Y_i' represents the production value in industry i after introducing the technology, namely the social benefits brought about by the introduced technology
P_{Ei} represents the payment cost of the enterprises' introduced technology
P_{Gi} represents the payment cost of government-introduced technology
λ_i represents the input–output elastic coefficient after introducing this technology

7.3.1.2 Target Analysis

Pursuit of interests is the basic demand of technology introduction. The enterprises that introduce technology hope to produce the maximum benefit by paying the lowest cost for technology introduction. Meanwhile, the introducers will not introduce the same technique twice. Therefore, the assumption is that there are n techniques for n industries to introduce in a certain period. The objective function is to maximize the social benefits in technology introduction. On the basis of Formula 7.8, we can get the objective function as shown in Formula 7.9.

$$\max Z = \sum_{i=1}^{n} A_i e^{\sigma_i t} \cdot L_i^{\alpha_i} \cdot K_i^{\beta_i} (P_{Ei} + P_{Gi})^{\lambda_i} \tag{7.9}$$

7.3.1.3 Constraint Conditions Recognition

The enterprises will introduce technology only when their own demands are satisfied. Therefore, the constrained condition of technology introduction is not taken into consideration here. Only the limitations of available resources, funds, systems, environmental bearing capacity, and other conditions are taken into consideration from the point of view of the local government during technology introduction.

1. Resource constraints

The introduction of technology needs certain ways, carriers, and channels. The introduced technology flows into this region and combines with various elements rationally and then produces benefits. Therefore, certain resources are required for this introduction and for the outcome of benefits, such as the allocation of education resources and the corresponding department of technology. This constraint condition is expressed as Formula 7.10.

$$\begin{gathered} \sum_{i=1}^{n} R_i \cdot (P_{Ei} + P_{Gi}) \le R_R \\ (R_R = R_{\text{Resource}}) \end{gathered} \tag{7.10}$$

where

R_i represents the resource consumed by unit cost after the introduction of the ith technique

R_R represents the quantity of resource that can be consumed

2. Capital constraints

Every year, the local government has a budget for expenditure, and therefore the investment in the introduction of technology cannot exceed the government expenditure budget. It is shown in Formula 7.11.

$$\sum_{i=1}^{n} P_{Gi} \le M_B \tag{7.11}$$

where

M_B represents the capital budget invested by the government in the introduction of technology during this period

3. System condition constraints

Whether the technology can be introduced rationally, allocated efficiently, and can play its role persistently is dependent on whether there is a system arrangement that is propitious to develop innovation activities and introduce technology. Generally system constraints include tax policy, distribution mechanism, security guarantee, interest returns, and so on. Moreover the patent system and industrial standardization system are also part of the constraints. This constraints condition is given by Formula 7.12.

$$\sum_{i=1}^{n} S_i \cdot (P_{Ei} + P_{Gi}) \leq S_R \tag{7.12}$$

where

S_i represents the transfer system guarantee needed by the unit cost after the transfer of the ith technology

S_R represents the local government's system constraints condition

4. Infrastructure capacity constraints

The efficiency and effect of the introduced technology also depends on the developing levels of local infrastructure which primarily includes infrastructure construction, technology market, education, and research institutions. This constraint condition is given by Formula 7.13.

$$\sum_{i=1}^{n} C_i \cdot (P_{Ei} + P_{Gi}) \leq C_F \tag{7.13}$$

where

C_i represents the infrastructure requirement of unit cost after the flow of the ith technology

C_F represents the infrastructure capacity in this region

7.3.1.4 Optimization Model of the Government Investment in Technology Introduction

Based on the analysis above, the optimization model of the government investment in technology introduction is shown in Formula 7.14.

$$\max Z = \sum_{i=1}^{n} A_i e^{\sigma_i t} \cdot L_i^{\alpha_i} \cdot K_i^{\beta_i} (P_{Ei} + P_{Gi})^{\lambda_i}$$

$$\begin{cases} \sum_{i=1}^{n} R_i \cdot (P_{Ei} + P_{Gi}) \leq R_R \\ \sum_{i=1}^{n} (P_{Ei} + P_{Gi}) \leq M_B \\ \sum_{i=1}^{n} S_i \cdot (P_{Ei} + P_{Gi}) \leq S_R \\ \sum_{i=1}^{n} C_i \cdot (P_{Ei} + P_{Gi}) \leq C_F \\ P_{Gi} \geq 0, \quad i = 1,2,\ldots,n \end{cases} \tag{7.14}$$

7.3.2 Study of the Properties of the Optimization Model of Government Investment in Technology Introduction

Based on the model above, the properties can be obtained as follows:

Proposition 7.9: During the same period, after increased investment of technology introduction in industry i in a certain region, assuming that the production value of industry i is a function of government investment, which is $Y_i' = f(P_{Gi})$, there are several conclusions:

1. $Y_i' = f(P_{Gi})$ has a continuous second derivative, and for arbitrary $P_{Gi} > 0$, $f'(P_{Gi}) > 0$.
2. $Y_i' = f(P_{Gi})$ is a strict concave function, namely $f''(P_{Gi}) < 0$. ■

Proof: (1) $f'(P_{Gi}) = A_i e^{\sigma_i t} \cdot L_i^{\alpha_i} \cdot K_i^{\beta_i} \cdot (P_{Ei} + P_{Gi})^{\lambda_i - 1} \cdot \lambda_i, \quad i = 1,2,\ldots,n$

So: for arbitrary $P_{Gi} > 0, f'(P_{Gi}) > 0$.

$$(2) \quad \frac{\partial^2 f(P_{Gi})}{\partial P_{Gi}^2} = A_i e^{\sigma_i t} \cdot L_i^{\alpha_i} \cdot K_i^{\beta_i} \cdot (P_{Ei} + P_{Gi})^{\lambda_{i-1}} \cdot \lambda_i \cdot (\lambda_i - 1), \quad i = 1,2,\ldots,n$$

So: $f''(P_{Gi}) < 0$.

In Proposition 7.9, $f'(P_{Gi}) > 0$ indicates that $Y_i' = f(P_{Gi})$ is strict monotone increasing function of P_{Gi}, $f''(P_{Gi}) < 0$ indicates that the increment of social benefits is decreasing step by step with the increase of government investment in industry i.

Proposition 7.10: The government investment optimization model in technology introduction is a convex programming model. ■

Taking a deformation of Model 7.14 Equation 7.15 can be deduced.

$$\min f(P_G) = -\sum_{i=1}^{n} A_i e^{\sigma_i t} \cdot L_i^{\alpha_i} \cdot K_i^{\beta_i} \cdot (P_{Ei} + P_{Gi})^{\lambda_i}$$

$$\begin{cases} R_R - \sum_{i=1}^{n} R_i \cdot (P_{Ei} + P_{Gi}) \geq 0 \\ M_B - \sum_{i=1}^{n} (P_{Ei} + P_{Gi}) \geq 0 \\ S_R - \sum_{i=1}^{n} S_i \cdot (P_{Ei} + P_{Gi}) \geq 0 \\ C_F - \sum_{i=1}^{n} C_i \cdot (P_{Ei} + P_{Gi}) \geq 0 \\ P_{Gi} \geq 0, \quad i = 1, 2, \ldots, n \end{cases} \tag{7.15}$$

For the model above, the condition to make it a convex function is that the objective function $f(P_G)$ is a convex function, and all the constraints are concave functions.

Proof: The objective function is taken as an example to prove it is a convex function.

The Hessian matrix of the objective function is

$$H(X^*) = \begin{bmatrix} \frac{\partial^2 f(X^*)}{\partial x_1^2}, \frac{\partial^2 f(X^*)}{\partial x_1 \partial x_2}, \ldots, \frac{\partial^2 f(X^*)}{\partial x_1 \partial x_5} \\ \frac{\partial^2 f(X^*)}{\partial x_2 \partial x_1}, \frac{\partial^2 f(X^*)}{\partial x_2^2}, \ldots, \frac{\partial^2 f(X^*)}{\partial x_2 \partial x_5} \\ \cdots \\ \frac{\partial^2 f(X^*)}{\partial x_5 \partial x_1}, \frac{\partial^2 f(X^*)}{\partial x_5 \partial x_2}, \ldots, \frac{\partial^2 f(X^*)}{\partial x_5^2} \end{bmatrix}$$

where

$$\frac{\partial^2 f(X^*)}{\partial x_i^2} = -A_i e^{\sigma_i t} \cdot L_i^{\alpha_i} \cdot K_i^{\beta_i} \cdot (P_{Ei} + x_i)^{\lambda_{i-1}} \cdot \lambda_i \cdot (\lambda_i - 1), \quad i = 1, 2, \ldots, 5, \quad 0 \leq \lambda_1 \leq 1$$

So

$$\frac{\partial^2 f(X^*)}{\partial x_i^2} \geq 0, \quad i = 1, 2, \ldots, 5,$$

because

$$\frac{\partial^2 f(X^*)}{\partial x_i \partial x_j} = 0, \quad i = 1, 2, \ldots, 5, \quad j = 1, 2, \ldots, 5, \quad i \neq j$$

So the Hessian matrix is positive semidefinite in X^*, and hence the objective function is convex. Therefore, the government investment optimization model in technology introduction is a convex programming model. ❑

Proposition 7.11: The government is certain to obtain the maximum social benefits by the introduction of technology. ■

Proof: For Formula 7.15, each constraint is introduced with generalized Lagrange multipliers respectively, which are $\gamma_1^*, \gamma_2^*, \ldots, \gamma_{n+4}^*$. Assume the point of K–T is P_G^*, so the K–T condition of this problem is shown as Formula 7.16:

$$\begin{cases} -\lambda_1 \cdot A_1 e^{\sigma_1 t} \cdot L_1^{\alpha_1} \cdot K_1^{\beta_1} \cdot (P_{E1} + P_{G1}^*)^{\lambda_1 - 1} + \gamma_1^* \cdot R_1 + \gamma_2 + \gamma_3^* \cdot S_1 + \gamma_4^* \cdot C_1 = 0 \\ -\lambda_2 \cdot A_2 e^{\sigma_1 t} \cdot L_1^{\alpha_2} \cdot K_1^{\beta_2} \cdot (P_{E2} + P_{G2}^*)^{\lambda_2 - 1} + \gamma_1^* \cdot R_2 + \gamma_2 + \gamma_3^* \cdot S_2 + \gamma_4^* \cdot C_2 = 0 \\ \ldots \\ -\lambda_n \cdot A_n e^{\sigma_n t} \cdot L_n^{\alpha_n} \cdot K_n^{\beta_n} \cdot (P_{En} + P_{Gn}^*)^{\lambda_n - 1} + \gamma_1^* \cdot R_n + \gamma_2 + \gamma_3^* \cdot S_n + \gamma_4^* \cdot C_n = 0 \\ \gamma_1^* \cdot \left[R_R - \sum_{i=1}^{n} R_i \cdot (P_{Ei} + P_{Gi}^*) \right] = 0 \\ \gamma_2^* \cdot \left[M_B - \sum_{i=1}^{n} P_{Gi}^* \right] = 0 \\ \gamma_3^* \cdot \left[S_R - \sum_{i=1}^{n} S_i \cdot (P_{Ei} + P_{Gi}^*) \right] = 0 \\ \gamma_4^* \cdot \left[C_F - \sum_{i=1}^{n} C_i \cdot (P_{Ei} + P_{Gi}^*) \right] = 0 \\ \gamma_i^* \geq 0, \quad i = 1, 2, 3, 4 \end{cases} \tag{7.16}$$

Because this nonlinear programming problem (Formula 7.15) is a convex program, the K–T condition is a necessary and sufficient condition to ensure the optimal point. If the K–T point P_G^* that meets the condition is obtained, it will be the local maximum point of the program. And the maximum social benefits can be obtained by distributing the funds for technology introduction based on the analysis above. ❑

Proposition 7.12: The investment coefficient of introduced technology for government is certain to be greater than 1. ■

Proof: Assuming that government's investment increases by ΔP, the added-value of social benefits is shown by Formula 7.17:

$$\sum_{i=1}^{n} A_i e^{\sigma_i t} \cdot L_i^{\alpha_i} \cdot K_i^{\beta_i} \cdot (P_{Ei} + P_{Gi} + \Delta P_i)^{\lambda_i} - \sum_{i=1}^{n} A_i e^{\sigma_i t} \cdot L_i^{\alpha_i} \cdot K_i^{\beta_i} \cdot (P_{Ei} + P_{Gi})^{\lambda_i}$$

$$= \sum_{i=1}^{n} A_i e^{\sigma_i t} \cdot L_i^{\alpha_i} \cdot K_i^{\beta_i} \cdot \Delta M \tag{7.17}$$

Assuming that the ratio of the social benefits added-value of certain technology obtained by increasing the investment of the added technology to the added-cost of government investment is expressed as σ, the investment coefficient for government introduced technology,

$$\sigma = \frac{A_i e^{\sigma_i t} \cdot L_i^{\alpha_i} \cdot K_i^{\beta_i} \cdot \Delta M}{\Delta P_i} \tag{7.18}$$

1. If $\sigma \leq 1$, for this piece of introduced technique, the added-investment cost of government is not enough to barter for social external spillover benefits, it is in contradiction with the purpose of introducing technology. At this time, the government does not invest more.
2. Only when $\sigma > 1$, the government will increase the investment in technology introduction. ❑

7.3.3 Break-Even Analysis of the Government Investment Model in Technology Introduction

Break-even analysis is an important method to carry out the economic analysis of production and management; it aims to find at what kind of production scale revenue and expenditure can reach a balance. The break-even analysis under the conditions of multifactor Cobb–Douglas production function are discussed and the break-even point of the smallest inputs from the government in different cases, for instance, unchanging, increasing, and decreasing returns to scale are designed, so as to look for an effective way of reducing costs and improving competitiveness.

7.3.3.1 Production Function

In Formula 7.7, in the same period, the production function of industry i after the technology investment is shown as follows:

$$Y_i' = A_i e^{\sigma_i t} \cdot L_i^{\alpha_i} \cdot K_i^{\beta_i} \cdot (P_{Ei} + P_{Gi})^{\lambda_i} \tag{7.19}$$

where

- Y_i' is the output value of industry i after the increase of technology investment, which is the social benefit for the increased investment
- P_{Ei} is the payment cost that the enterprises incur for technology investment
- P_{Gi} is the expense that the government incurs for the positive external effects of this imported technology, in short, government expenditure
- λ_i is the input–output elasticity coefficient after technology investment is enhanced

The production function reflects the relationship between input and output. It demonstrates the relationship between the input and the maximum output produced, or between the established production and the minimum input required. In the following discussion, the production function after n industries have increased technology investment in a region on the basis of the same period is considered. This function is shown in Formula 7.20.

$$Y_i' = \prod_{i=1}^{n} A_i e^{\sigma_i t} \cdot L_i^{\alpha_i} \cdot K_i^{\beta_i} \cdot (P_{Ei} + P_{Gi})^{\lambda_i} \tag{7.20}$$

According to the definition of output elasticity of factors, it is easy to conclude that the output elasticity of P_{Gi} is as follows. Here P_{Gi} represents the government technique investment in industry i.

$$E_{P_{Gi}} = \frac{\partial Y'}{\partial P_{Gi}} \frac{P_{Gi}}{Y'} = \lambda_i, \quad i = 1, 2, \ldots, n \tag{7.21}$$

It demonstrates that the parameter λ_i has a clear economic significance in the production function $Y_i' = A_i e^{\sigma_i t} \cdot L_i^{\alpha_i} \cdot K_i^{\beta_i} \cdot (P_{Ei} + P_{Gi})^{\lambda_i}$, and $A_i e^{\sigma_i t}$ in the production function $Y_i' = A_i e^{\sigma_i t} \cdot L_i^{\alpha_i} \cdot K_i^{\beta_i} \cdot (P_{Ei} + P_{Gi})^{\lambda_i}$ is the coefficient of efficiency, and reflects the response of the broad technology progress levels; thus $0 \le \lambda_i \le 1, (i = 1,2,\ldots,n)$, $A_i e^{\sigma_i t} > 0$. Let $\lambda = \sum_{i=1}^{n} \lambda_i$, then λ is the well known scale elasticity. The production function is the λ step homogeneous function. When $\lambda = 1$, the production is the unchanging returns to scale; When $\lambda > 1$, the production is the increasing returns to scale; when $\lambda < 1$, it is the declining returns to scale.

7.3.3.2 Break-Even Analysis with Consideration of Production Function

The break-even analysis in the general sense is based on comprehensive analysis of the traffic (production, sale quantity, or sale volume), costs, and profits which impose constraints on each other. It can be used to make a prior decision on the level of production or sales to avoid the deficit thus providing information for business decision-making. The break-even analysis of production function considered in this paper has further demands based on the origin that requires the government to use the factors of production efficiently by intensive operation, and explore ways to minimize the cost of the technology inputs to break even. The following discussion is based on mathematical models.

It is assumed that $P_{Gi} = (P_{G1}, P_{G2}, \ldots, P_{Gn})^T$ is the vector of government's technology input and the amount of the input is denoted by F, which can also be taken as fixed cost. If only the government's investment is considered, that is $P_{Ei} = 0$, then the break-even balance model of production function is given by Formula 7.22:

$$\min TC = \sum_{i=1}^{n} P_{Gi}$$
$$\text{s.t.} \begin{cases} \prod_{i=1}^{n} A_i e^{\sigma_i t} \cdot L_i^{\alpha_i} \cdot K_i^{\beta_i} P_{Gi}^{\lambda_i} - \sum_{i=1}^{n} P_{Gi} = F \\ P_{Gi} \ge 0, \quad (i = 1,2,\ldots,n) \end{cases} \tag{7.22}$$

Using the Lagrange Multiplier Method to solve the model, and constructing the Lagrange function:

$$L(P_{G1}, P_{G2}, \ldots, P_{Gn}, \alpha) = \sum_{i=1}^{n} P_{Gi} + \alpha\left[\prod_{i=1}^{n} A_i e^{\sigma_i t} \cdot L_i^{\alpha_i} \cdot K_i^{\beta_i} P_{Gi}^{\lambda_i} - \sum_{i=1}^{n} P_{Gi} - F\right] \tag{7.23}$$

Solving the partial derivative, and equating the partial derivative to 0, then:

$$\begin{cases} \dfrac{\partial L}{\partial P_{Gi}} = 1 + \alpha A_i e^{\sigma_i t} \cdot L_i^{\alpha_i} \cdot K_i^{\beta_i} \cdot P_{Gi}^{\lambda_i - 1} \cdot \lambda_i - \alpha = 0 \\ \dfrac{\partial L}{\partial \alpha} = \prod_{i=1}^{n} A_i e^{\sigma_i t} \cdot L_i^{\alpha_i} \cdot K_i^{\beta_i} P_{Gi}^{\lambda_i} - \sum_{i=1}^{n} P_{Gi} - F = 0 \end{cases} \quad i = 1,2,\ldots,n \tag{7.24}$$

This is the necessary condition of Lagrange function for reaching the extreme value, which can be further simplified as follows:

$$\prod_{i=1}^{n} A_i e^{\sigma_i t} \cdot L_i^{\alpha_i} \cdot K_i^{\beta_i} P_{Gi}^{\lambda_i} = \frac{(\alpha-1)P_{Gi}}{\alpha\lambda_i}, \quad i=1,2,\ldots,n$$

$$\prod_{i=1}^{n} A_i e^{\sigma_i t} \cdot L_i^{\alpha_i} \cdot K_i^{\beta_i} P_{Gi}^{\lambda_i} = \sum_{i=1}^{n} P_{Gi} + F$$

The following equations are generated:

$$\begin{cases} \dfrac{(\alpha-1)P_{Gi}}{\alpha\lambda_i} = \sum\limits_{i=1}^{n} P_{Gi} + F \\ \dfrac{P_{Gi}}{\lambda_i} = \dfrac{P_{Gj}}{\lambda_j} \end{cases} \quad i,j=1,2,\ldots,n \tag{7.25}$$

Hence

$$P_{Gi} = F\alpha\lambda_i/[\alpha(1-\lambda)-1] \tag{7.26}$$

where, $\lambda = \sum_{i=1}^{n} \lambda_i$

Substitute Formula 7.26 in Equation 7.24 and simplifying the equation:

$$\prod_{i=1}^{n} A_i e^{\sigma_i t} \cdot L_i^{\alpha_i} \cdot K_i^{\beta_i} \lambda_i^{\lambda_i} \left[\frac{F\alpha}{\alpha(1-\lambda)-1}\right]^{\lambda} = \frac{F(\alpha-1)}{\alpha(1-\lambda)-1}$$

Let $\prod_{i=1}^{n} A_i e^{\sigma_i t} \cdot L_i^{\alpha_i} \cdot K_i^{\beta_i} \lambda_i^{\lambda_i} = M$, then:

$$M\left[\frac{F\alpha}{\alpha(1-\lambda)-1}\right]^{\lambda} = \frac{F(\alpha-1)}{\alpha(1-\lambda)-1} \tag{7.27}$$

According to the implicit function theorem, the implicit function $\alpha = \varphi(\lambda)$ can be determined based on Formula 7.27, and substituting it in Formula 7.26, then:

$$P_{Gi} = \frac{F\lambda_i\varphi(\lambda)}{[\varphi(\lambda)(1-\lambda)-1]}, \quad i=1,2,\ldots,n \tag{7.28}$$

Formula 7.28 is the solution of Model 7.22. The solution under different values of λ are discussed below.

1. When $\lambda = 1$, Formula 7.27 turns to

$$MF\alpha = F(\alpha - 1)$$

$$\alpha = \frac{1}{1-M} = \frac{1}{1-\prod_{i=1}^{n} A_i e^{\sigma_i t} \cdot L_i^{\alpha_i} \cdot K_i^{\beta_i} \lambda_i^{\lambda_i}} \tag{7.29}$$

Substitute Formula 7.29 in Formula 7.26, then:

$$P_{Gi} = \frac{F\lambda_i}{\left[\prod_{i=1}^{n} A_i e^{\sigma_i t} \cdot L_i^{\alpha_i} \cdot K_i^{\beta_i} \lambda_i^{\lambda_i} - 1\right]}, \quad i = 1,2,\ldots,n \tag{7.30}$$

It can be proved that Formula 7.30 is the whole optimal solution of Model 7.22, that is, when production is in the unchanging returns to scale, Formula 7.30 gives the best use of production factors in the break-even balance, in this case, the minimum cost, which is

$$TC^* = \sum_{i=1}^{n} \frac{F\lambda_i}{\prod_{i=1}^{n} A_i e^{\sigma_i t} \cdot L_i^{\alpha_i} \cdot K_i^{\beta_i} \lambda_i^{\lambda_i} - 1} = \frac{F}{\prod_{i=1}^{n} A_i e^{\sigma_i t} \cdot L_i^{\alpha_i} \cdot K_i^{\beta_i} \lambda_i^{\lambda_i} - 1}$$

2. When $\lambda \neq 1$, making Formula 7.2 meaningful requires that it should satisfy the following conditions:

$$\frac{(\alpha-1)}{[\alpha(1-\lambda)-1]} > 0, \quad \frac{(\alpha-1)}{[\alpha(1-\lambda)-1]} > 0 \tag{7.31}$$

To further illustrate the meaning of α, by Equation 7.24, it is clear that:

$$\alpha = \frac{1}{[1-(\partial L/\partial P_{Gi})]}, \quad i = 1,2,\ldots,n \tag{7.32}$$

where $\partial L/\partial P_{Gi}$ stands for the marginal productivity of the input element P_{Gi}. It is clear from Formula 7.32 that when the marginal productivity of P_{Gi} is more than P_{Gi}, then $\alpha > 0$, whereas when the marginal productivity of P_{Gi} is less than P_{Gi}, $\alpha < 0$. So the symbol of α value stands for the marginal construction effect of government technology investment.

Solving the system of Inequalities 7.31 under two conditions as follows:

a. When $\lambda > 1$, the solution of Inequalities 7.31 is

$$1/(1-\lambda) < \alpha < 0 \tag{7.33}$$

The range of the α value shows that, when the production belongs to the increasing returns to scale, the effect of production and management is balanced. The more α value is close to zero, the better the production and management results.

b. When $\lambda < 1$, the solution of Inequalities 7.31 is

$$\alpha > 1/(1-\lambda) > 0 \quad \text{or} \quad \alpha < 0 \tag{7.34}$$

This shows that, when production belongs to the declining returns to scale, the marginal productivity of elements may be lower than the price of elements, and this leads to the decline of economic benefit.

7.3.4 Case Study

7.3.4.1 Government Investment and Planning Model in Technology Introduction

Assume that five kinds of technologies are introduced in five different industries respectively, so $n = 5$. The indicator values of the model which are based on calculation and experience are shown in Tables 7.7 and 7.8.

According to the conditions above, the optimization model of government investment in technology introduction can be established as in Equation 7.35.

$$\max Z = 938.997 \times (50 + x_1)^{0.3} + 558.74 \times (40 + x_2)^{0.35} + 868.02 \times (35 + x_3)^{0.25}$$
$$+ 533.33 \times (40 + x_4)^{0.35} + 1076.73 \times (30 + x_5)^{0.2}$$

$$\text{s.t.} \begin{cases} 0.8 \times x_1 + 0.6 \times x_2 + 0.45 \times x_3 + 0.5 \times x_4 + 0.5 \times x_5 \le 885.25 \\ x_1 + x_2 + x_3 + x_4 + x_5 \le 1200 \\ 0.3 \times x_1 + 0.35 \times x_2 + 0.4 \times x_3 + 0.25 \times x_4 + 0.2 \times x_5 \le 441 \\ 0.65 \times x_1 + 0.5 \times x_2 + 0.65 \times x_3 + 0.4 \times x_4 + 0.45 \times x_5 \le 595.25 \\ x_i \ge 0, \quad i = 1, 2, \ldots, 5 \end{cases} \tag{7.35}$$

Employing software lingo, results shown in Table 7.9 can be obtained.

Table 7.7 Indicators of Production Function and Unit Resource Consumption in Different Industries

Technology i	*1*	*2*	*3*	*4*	*5*
A	10	8	5	7	4
σ	0.004	0.007	0.007	0.004	0.001
T	1	1	1	1	1
L	500	400	300	400	300
K	800	600	500	500	400
α	0.3	0.35	0.4	0.3	0.4
β	0.4	0.3	0.35	0.35	0.4
λ	0.3	0.35	0.25	0.35	0.2
P	50	40	35	40	30
R_i	0.8	0.6	0.45	0.5	0.5
S_i	0.3	0.35	0.4	0.25	0.2
E_i	0.65	0.5	0.65	0.4	0.45

Note: Fund unit, million; labor unit, million people.

Table 7.8 Constraint Conditions of Regional Resources (Unit: Million)

Quantity of Resources the Region Could Provide	*Capital Budget*	*Constraint*	*Capacity of Infrastructure*
1000	1200	500	700

Table 7.9 Result of Calculation

Objective Value					*19917.54*
Variable	*Value*	*Reduced Cost*	*Row*	*Slack or Surplus*	*Dual Price*
X_1	222.8149	0.000000	1	19917.54	1.000000
X_2	318.6593	$-0.2791587 \times 10^{-7}$	2	195.0008	0.000000
X_3	97.55292	0.6616175×10^{-7}	3	7.167131	0.000000
X_4	430.6265	0.000000	4	91.31113	0.000000
X_5	123.1792	$-0.3600319 \times 10^{-7}$	5	0.000000	8.546029

7.3.4.2 Result Analysis

1. In this case, the infrastructure ability of the region is the compact constraint which limits the region from achieving higher social benefits. Hence if the government wants to achieve more social benefits from the technology imported, enhancement of the region's infrastructure is required.
2. A full allocation of the government funds shows that technology introduction has a positive role in increasing social profits and promoting the development of the region. It also demonstrates that the government holds a positive attitude to the investment in technology introduction.
3. Larger input elasticity of the government cost indicates greater influence that the government investment has on the output. The profits will significantly increase if government investment is increased, so the government investment in technology introduction in this industry will increase.

Chapter 8

Game Strategy Analysis on Technology Transfer and Technology Innovation

In recent years, many scholars have carried out research on technology innovation, technology diffusion, technology spillover, and other issues of technology transfer, using Game Theory. On the basis of some fundamental hypotheses, this section will discuss the application of game theory analysis to these issues. By looking deeper, a virtual circle strategy for technology transfer, which can provide a reference for market construction and institutional arrangements and provide a new vision for the study of technology transfer, can be evolved.

8.1 Game Model and Strategy Analysis of Technology Innovation

Technology innovation plays a very important part in the process of technology transfer. If there is no technology innovation, there will be no technology transfer, technology diffusion, technology spillover, and hence, a series of technology transfers. In today's society, corporations are in increasingly competitive environments, and there have been many large monopoly enterprises whose innovations have had a significant impact on technology progress.

8.1.1 Static Game Analysis with Complete Information on Duopoly Enterprise's Technology Innovation

Since D'Aspremont and Jacque Minde (1988) made use of a duopoly game model to study the technology innovation for the first time, Game Theory has been used by many scholars to dispose off issues of technology innovation. Kotaro Suzumura (1992) established a two-stage dynamic game model for technology innovation, and Nicholas S. Vonotras (1989) put forward a three-stage game model; then Ziss (1944) also built a spillover two-stage R&D bilateral oligopoly game model. But all of these researchers based their strategies on the output of enterprises, without considering the technical content. Therefore, this chapter will study technology innovation of duopoly enterprises taking the actual situation into consideration.

8.1.1.1 Construction of the Game Model

It is assumed that two enterprises, in a balance of power, will innovate for the same project, and manufacture products with the same quality. Furthermore, each of them will choose its own action strategies based successively on the same information, aiming at maximum profit. The reason for the assumption of complete information is that because they would have had many instances of competition and cooperation and would have a good understanding of each other on their core competencies, the quality of personnel, and other basic situations. First, the following hypotheses are made:

Hypothesis 8.1: When the product technology is improved, the production costs decrease without change in quality. The market-clearing price p is an antidemand function of the output Q, that is, when the demand of Q is certain, production price is $p(Q)=a-bQ$. (for the convenience of calculation, $b=1$ is assumed). Q is the total output of the market, which is also the total output of the two monopolies, $Q=q_1+q_2$, q_1 is the output of Enterprise 1, q_2 is the output of Enterprise 2, and a represents market scale. ■

Hypothesis 8.2: The technical content of both Enterprises 1 and 2 after independent technology innovation is t_1 and their technical content after cooperative technology innovation is t_2. The returns, from the investment in technology innovation, which is a quadratic function of technical content t_i $(t_i>0)$, diminish gradually. Usually the higher the technical content t_i is, the greater the market attraction will be. Also, innovation investment is a convex function of t_i, so we can define the input function of technology innovation as follows:

$$W_i = \frac{1}{2}\beta t_i^2 \quad (i=1,2)$$

such that $\beta > 0$ and β is an input parameter of marginal technical content.

This shows that the investment unit increases along with the technical content, which is in line with the actual process of innovation.

$$c_i = C - kt_i \quad (i = 1,2 \quad kt_i < C)$$ ■

Hypothesis 8.3: Both Enterprises 1 and 2 have no fixed cost, and their unit product cost before innovation has a fixed value C $(C<a)$. After innovation the product cost is a decreasing function of technical content; that means the higher the technical content, the lower the cost. If it is assumed that the cost decreasing coefficient of unit product technical content is k, then the cost function of the unit product can be defined as

$$c_i = C - kt_i \quad (i = 1,2 \quad kt_i < C)$$ ■

Proposition 8.1: If the Hypotheses 8.1, 8.2, and 8.3 hold good, the profit of Enterprise i is as follows:

$$\pi_i = q_i(p(Q) - c_i) - \frac{1}{2}\beta t_i^2 \quad (i = 1,2)$$ ■

Proof: By Hypothesis 8.1 it is clear that the total income R_i of Enterprise i equals the product of production q_i and price $p(Q)$.

$$R_i = q_i^* \, p(Q)$$

By the Hypotheses 8.2 and 8.3 it is clear that the total cost Y_i of the Enterprise i equals the sum of technology innovation investment and the cost of the product.

$$Y_i = q_i^* \, c_i + \frac{1}{2}\beta t_i^2$$

As a result, the corporate profit π_i equals the difference between its total revenue R_i and its total cost Y_i,

$$\pi_i = q_i^* \, c_i + \frac{1}{2}\beta t_i^2 = q_i(p(Q) - c_i) - \frac{1}{2}\beta t_i^2$$ ❑

8.1.1.2 Solution and Analysis of Game Model

1. Solution and analysis of game model of the enterprise carrying out independent technology innovation

By Proposition 8.1, while both Enterprises 1 and 2 both have carried out an independent technology innovation and the profit of Enterprise 1 is

$$\pi_1 = q_1(p(Q) - c_1) - \frac{1}{2}\beta t_1^2$$

$$= q_1(a - (q_1 + q_2) - C + kt_1) - \frac{1}{2}\beta t_1^2 \tag{8.1}$$

and the profit of Enterprise 2 is

$$\pi_2 = q_2(p(Q) - c_1) - \frac{1}{2}\beta t_1^2$$

$$= q_2(a - (q_1 + q_2) - C + kt_1) - \frac{1}{2}\beta t_1^2 \tag{8.2}$$

If the two sides maximize their profit, then

$$\max_{q_1} \pi_1 (q_1, q_2) = \max(q_1^* (a - (q_1 + q_2) - C + kt_1) - \frac{1}{2}\beta t_1^2$$

$$\max_{q_2} \pi_1 (q_1, q_2) = \max(q_2^* (a - (q_1 + q_2) - C + kt_1) - \frac{1}{2}\beta t_1^2$$

$$\text{s.t.} \quad q_1 \geq 0, \quad q_2 \geq 0, \quad a > C, \quad k > 0, \quad \beta > 0, \quad t_1 > 0$$

Proposition 8.2: If Proposition 8.1 holds good, the Cournot–Nash equilibrium output of the enterprise is given by

$$q_1^* = q_2^* = \frac{a - C + kt_1}{3}$$

■

Proof: The way to obtain the Cournot–Nash equilibrium output is to find the first derivative of the enterprise's profit function and equate it to 0. The following formulas can then be deduced from Formulas 8.1 and 8.2:

$$\frac{\partial \pi_1}{\partial q_1} = a - (q_1 + q_2) - C + kt_1 - q_1 = 0 \tag{8.3}$$

$$\frac{\partial \pi_2}{\partial q_2} = a - (q_1 + q_2) - C + kt_1 - q_2 = 0 \tag{8.4}$$

The solution to the equation group above, gives the Cournot–Nash equilibrium output.

$$q_1^* = q_2^* = \frac{a - C + kt_1}{3} \tag{8.5}$$

❑

In Formula 8.5, q_1^*, q_2^* are both increasing functions of a, t_1, k. It shows that the larger the market size, the higher the technical content of technology innovation. After innovation, the lower the unit product cost, the more the production.

Proposition 8.3: If Proposition 8.2 holds good, the Cournot–Nash equilibrium profit of the enterprise is as follows:

$$\pi_1^* = \pi_2^* = \left(\frac{a - C + kt_1}{3}\right)^2 - \frac{1}{2}\beta t_1^2$$

■

Proof: Using the result of Formula 8.5 in Formulas 8.1 and 8.2, the Cournot–Nash equilibrium profit of the two enterprises is as follows:

$$\pi_1^* = \pi_2^* = \left(\frac{a - C + kt_1}{3}\right)^2 - \frac{1}{2}\beta t_1^2 \tag{8.6}$$

From Formula 8.6, it is clear that the equilibrium profit of the enterprise is a quadratic function of the cost decreased coefficient k of unit product technical content, that is, after technology innovation, the larger the cost decreased coefficient k, the more the enterprise's profit. ❑

Proposition 8.4: If Proposition 8.3 holds good, after the independent innovation of the two enterprises, the best technical content is given by

$$t_1^* = \frac{2k(a - C)}{9\beta - 2k^2}$$

■

Proof: Differentiating 8.6 with respect to t_1 and equating the derivative to 0, it follows that:

$$\frac{\partial \pi_1^*}{\partial t_1} = 2\frac{a - C + kt_1}{3} * \frac{k}{3} - \beta t_1 = 0 \tag{8.7}$$

Therefore

$$t_1^* = \frac{2k(a-C)}{9\beta - 2k^2} \tag{8.8}$$

Because

$$k > 0, \quad a > C, \quad \beta > 0, \quad t_1 > 0,$$

Hence

$$9\beta - 2k^2 > 0$$

For

$$\frac{\partial^2 \pi_1^*}{\partial t_1^2} = \frac{2k^2 - 9\beta}{9}$$

So

$$\frac{\partial^2 \pi_1^*}{\partial t_1^2} < 0$$

Therefore, at the value of t_1^*, the function reaches the maximum value. The two enterprises have carried out independent innovations, the best technical content is available, and it follows that:

$$t_1^* = \frac{2k(a-C)}{9\beta - 2k^2}$$

Proposition 8.4 proves that technology innovation does not mean that the higher the technical content, the better the earnings, but when technical content equals to t_1^*, innovation profit is the greatest. In other words, when the innovation attains a certain degree, it needs to improve steadily with the advancement of technology. Any attempt to speed up the advancement in technology will require too much investment and this is likely to exceed the income. This situation is similar to what happens in some industries in reality.

2. Comparison between enterprises involved in independent technology innovation and cooperative technology innovation

Proposition 8.5: If Hypotheses 8.1, 8.2, and 8.3 hold good, the best output of cooperative technology innovation is

$$Q^* = \frac{a - C + kt_2}{2}$$

and the largest profit is

$$\pi'^* = \left(\frac{a - C + kt_2}{2}\right)^2 - \frac{1}{2}\beta t_2^2$$

and the best technical content is

$$t_2^* = \frac{k(a - C)}{2\beta - k^2}$$

■

Proof: By Hypotheses 8.1, 8.2, and 8.3, it is clear that the profit of cooperative technology innovation is as follows:

$$\pi' = Q(p(Q) - c_2) - \frac{1}{2}\beta t_2^2 \tag{8.9}$$

Differentiating Formula 8.9 with respect to Q and equating it to 0, it follows that:

$$\frac{\partial \pi'}{\partial Q} = a - 2Q - C + kt_2 = 0 \tag{8.10}$$

Therefore the best output of cooperative technology innovation is as follows:

$$Q^* = \frac{a - C + kt_2}{2} \tag{8.11}$$

Applying the value of Formula 8.11 in Formula 8.9, the greatest profit of cooperative technology innovation can be deduced as follows:

$$\pi'^* = \left(\frac{a - C + kt_2}{2}\right)^2 - \frac{1}{2}\beta t_2^2$$

$$\frac{\partial \pi'^{*}}{\partial t_2} = 2\frac{a - C + kt_2}{2} * \frac{k}{2} - \beta t_2 = 0$$

So

$$t_2^* = \frac{k(a - C)}{2\beta - k^2} \tag{8.12}$$

❑

Proposition 8.6: If Proposition 8.5 holds good, the total output Q^* of cooperative technology innovation is less than that of the independent technology innovation, and the total profit of cooperative technology innovation is more than that of the independent technology innovation, and the best technical content of cooperative technology innovation t_1^* is larger than that of the independent technology innovation t_2^*, it follows that:

$$Q^* < q_1^* + q_2^*$$

$$\pi'^* > \pi_1^* + \pi_2^*$$

$$t_1^* < t_2^*$$

■

Proof: From Proposition 8.2, the total output of the two enterprises that have independent technology innovation is shown as follows:

$$q_1^* + q_2^* = 2\frac{a - C + kt_1}{3} \tag{8.13}$$

From Proposition 8.3, it is clear that the total profit of the two enterprises that have independent technology innovation is as follows:

$$\pi_1^* + \pi_2^* = 2\left(\frac{a - C + kt_1}{3}\right)^2 - \beta t_1^2 \tag{8.14}$$

From Formulas 8.13 and 8.14, and Propositions 8.4 and 8.5, we obtain that:

$$Q^* < q_1^* + q_2^*$$

$$\pi'^* > \pi_1^* + \pi_2^*$$

$$t_1^* < t_2^*$$

❑

Through the analysis above, it is clear that if the technical content of cooperative technology innovation is the same as that of the independent technology innovation, the total output of the two enterprises that have independent technology innovation is more than that of the cooperative technology innovation, and the total profit of the two enterprises that have independent technology innovation is less than that of the cooperative technology innovation, as they only take their own profit into account and neglect the external negative effect to the other enterprise, which is a typical Prisoners Dilemma. If the two sides collude, they are able to achieve more profit through higher output, thus getting out of the Prisoners Dilemma.

Moreover, $t_1^* < t_2^*$ implies that under the same resource condition, the best technical content of cooperative technology innovation is greater than that of independent technology innovation. This implies that, with other terms remaining unchanged, cooperative innovation can increase the technical content being achieved to the maximum. This is incompatible with the result obtained by taking merely the output into decision-making. To the whole society, from the production point of view, noncooperation is efficient, because it increases the output and lowers the prices. But from the technology point of view, or in terms of social progress, cooperative innovation is more favorable to technology progress. So instead of calling for competition blindly, a long-term view should be taken, encouraging cooperation between enterprises, especially those that have a lot of scope for innovation. Nevertheless, cooperation or competition, the best technical content of innovation is inversely proportionate to β; the bigger β is, the lower the best technical content achieved.

8.1.1.3 Numerical Analysis

1. Impact of product technical content on equilibrium profit

Assuming that $\alpha = 200$, $C = 10$, $k = 1$, $\beta = 100$, and technical content rises from 0.01 in steps of 0.02, from Proposition 8.3, the impact of product technical content on equilibrium profit, which is shown in Figure 8.1 can be obtained.

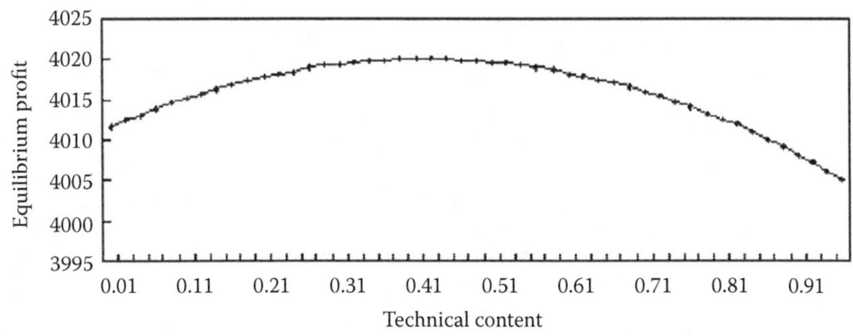

Figure 8.1 The impact of product technical content on equilibrium profit.

As can be seen from Figure 8.1, equilibrium profit of the enterprise improves along with the increase of technical content at first, but declines after a period of time, as the foregoing conclusion shows.

2. Impact of cost decreased coefficient of unit product technical content on the best technical content

Assuming that $\alpha = 200$, $C = 10$, $\beta = 100$, and unit product technical content's cost decreased coefficient k rises from 0.01 in steps of 0.04, by Proposition 8.4 the impact of unit product technical content's cost decreased coefficient on the best technical content can be obtained as shown in Figure 8.2.

As can be seen from Figure 8.2, the best technical content is directly proportionate to unit product technical content's cost decreased coefficient k. If other conditions remain unchanged, larger k means that greater technical content can be achieved. This is understandable, as after innovation, the lower the cost, the more the enterprises' profit, and then it is more beneficial for them to increase investment, so they can get a higher level of the best technical content.

3. Impact of marginal technical content's input parameter β on the best technical content

Assuming that $\alpha = 200$, $C = 10$, $k = 1$, and the marginal technical content's input parameter β rises from 50 in steps of 5, by Proposition 8.4 the impact of the marginal technical content's input parameter β on the best technical content can be obtained as shown in Figure 8.3.

As can be seen from Figure 8.3, the best technical content is inversely proportinate to the marginal technical content's input parameter β. A larger β indicates higher barriers to entry into the industry, which makes the technology of this industry hard to imitate. Meanwhile, to get a higher return, the enterprise of this industry has to increase the technical content of its products and continuously

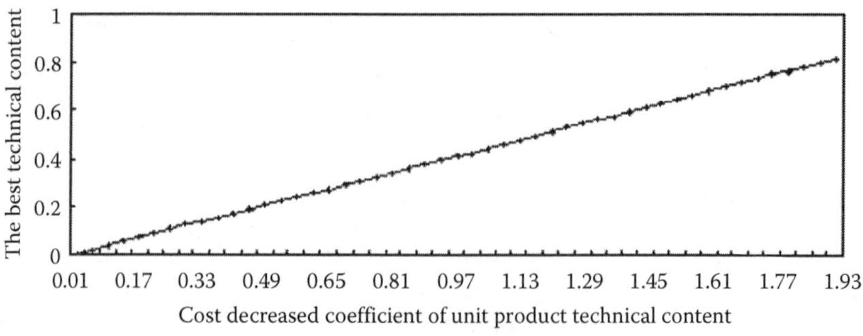

Figure 8.2 Cost decreased coefficient of unit product technical content and the best technical content.

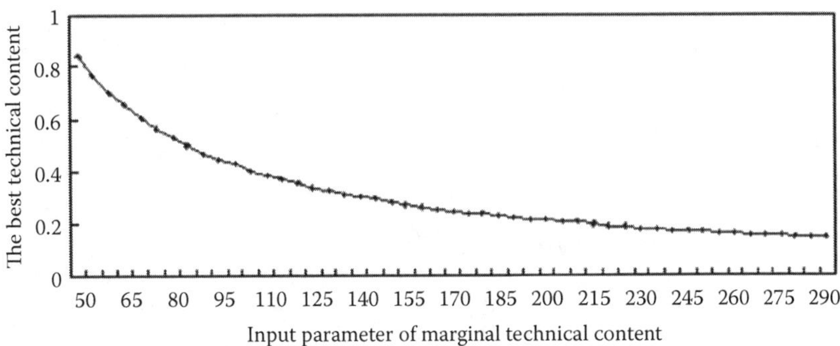

Figure 8.3 The best technical content and input parameter of marginal technical content.

carry out technical innovation, so that it can keep the barriers to entry advantage. And technology innovation needs more technology investment. The enterprises in the high-tech industry usually practice cooperation or merger to make larger investments. As a result, it is easier for high-technology industry to form an oligopoly.

8.1.2 Dynamic Game Analysis of Complete Information Theory of Technology Innovation

In the static game model of complete information theory it is assumed that two enterprises innovate and produce homogeneous products. However, in real life, it is unlikely for two enterprises to take action at the same time. Generally, one enterprise always takes action first and the other operates after it observes the first one, which forms a subgame Refinement Nash Equilibrium.

One enterprise takes action first and the other operates after it observes, and each of them targets profit maximization after deducting the cost. Enterprise 1's profit is $\pi_1 = (p(Q) - c_1)Q$ without innovation, but $\pi_1 = (p(Q) - c_1)Q - \frac{1}{2}\beta t_1^2$ after innovation; Enterprise 2 observes Enterprise 1 to decide whether it should innovate and then choose its own. As a result, it is a Stacklberg game.

8.1.2.1 Game Model of Enterprises 1 and 2 with Equal Technical Content after Innovation

1. Modeling and analysis

Hypothesis 8.4: Enterprise 1 innovates first, and its output is q_1. Enterprise 2 innovates later, and its output is q_2. We calculate from back to front, making use of backward induction. ■

Proposition 8.7: If the Hypotheses of 8.1, 8.2, 8.3, and 8.4 hold good, equilibrium output of Enterprise 1 is $q_1^* = \frac{a-C+kt_1}{2}$, whereas that of Enterprise 2 is $q_2^* = \frac{a-C+kt_1}{4}$. So the total output of the two enterprises is as follows:

$$Q = q_1^* + q_2^* = \frac{3(a-C+kt_1)}{4}$$

■

Proof: As can be seen from the Hypotheses of 8.1, 8.2, 8.3, and 8.4, the 's profits of Enterprise 2 are as follows:

$$\begin{aligned}\pi_2 &= q_2(p(Q)-c_1)-\frac{1}{2}\beta t_1^2 \\ &= q_2(a-(q_1+q_2)-C+kt_1)-\frac{1}{2}\beta t_1^2\end{aligned} \tag{8.15}$$

Because Enterprise 1 makes its decision earlier, Enterprise 2 is aware of its output. As a result, Enterprise 2 will make $\frac{\partial \pi_2}{\partial q_2}=0$ considering q_1 to be decided. Its best output can be obtained as follows:

$$q_2^* = \frac{a-q_1-C+kt_1}{2} \tag{8.16}$$

As Enterprise 1 can serve as a forecast for Enterprise 2 to choose q_2^*, its profit function is as follows:

$$\begin{aligned}\pi_1 &= q_1(a-(q_1+q_2^*)-C+kt_1)-\frac{1}{2}\beta t_1^2 \\ &= q_1(a-q_1-\frac{a-q_1-C+kt_1}{2}-C+kt_1)-\frac{1}{2}\beta t_1^2\end{aligned} \tag{8.17}$$

Let

$$\frac{\partial \pi_1}{\partial q_1} = \frac{a-2q_1-C+kt_1}{2} = 0$$

The equilibrium output of Enterprise 1 is as follows:

$$q_1^* = \frac{a - C + kt_1}{2} \tag{8.18}$$

Using Formula 8.18 in Formula 8.16, the equilibrium output of Enterprise 2 is obtained as follows:

$$q_2^* = \frac{a - C + kt_1}{4} \tag{8.19}$$

As can be seen from the Formulas 8.18 and 8.19, the total output of the two enterprises is as follows:

$$Q = q_1^* + q_2^* = \frac{3(a - C + kt_1)}{4} \tag{8.20}$$

❑

Add

$$\frac{\partial q_1^*}{\partial t_1} = \frac{k}{2}$$

$$\frac{\partial q_2^*}{\partial t_1} = \frac{k}{4}$$

The output rate of change due to technical content has no relation with anything but the cost decreased coefficient of unit product technical content, that is, the larger the cost decreased coefficient of unit product technical content, the higher the output. If this cost decreased coefficient is constant, the technical content's impact on output is unchanged, but the output of the enterprise changes more when technical content increases too much. In other words, in a dynamic game model, there is a first-mover advantage, as it is easier for the enterprise which has innovated earlier to seize the market, create brands, and then share the market.

Proposition 8.8: If Proposition 8.1 holds good, profits of Enterprises 1 and 2 are as follows:

$$\pi_1^* = \frac{(a - C + kt_1)^2}{8} - \frac{1}{2}\beta t_1^2$$

$$\pi_2^* = \left(\frac{a - C + kt_1}{4}\right)^2 - \frac{1}{2}\beta t_1^2$$

■

Proof: Using Formulas 8.18 and 8.19 in Formulas 8.16 and 8.17, the profits of Enterprises 1 and 2 can be obtained as follows:

$$\pi_1^* = \frac{(a - C + kt_1)^2}{8} - \frac{1}{2}\beta t_1^2$$

$$\pi_2^* = \left(\frac{a - C + kt_1}{4}\right)^2 - \frac{1}{2}\beta t_1^2$$

In Cournot–Nash equilibrium, the total output of the two enterprises after innovation is

$$Q = q_1^* + q_2^* = \frac{2(a - C + kt_1)}{3}$$

In the Stacklberg game model, the total output from backward induction is

$$\frac{3(a - C + kt_1)}{4}$$

which is higher than that in Cournot–Nash equilibrium. Thereby the market-clearing price in the Stacklberg game model is lower. Enterprise 1's profit in the Stacklberg game model is

$$\pi_1^* = \frac{(a - C + kt_1)^2}{8} - \frac{1}{2}\beta t_1^2$$

which is higher than $\left(\frac{a - C + kt_1}{3}\right)^2 - \frac{1}{2}\beta t_1^2$ in the Cournot–Nash equilibrium. In contrast, the profit of Enterprise 2 in the Stacklberg game model decreases, which shows that a participant who is handed more information advantage is also likely to suffer damage.

Proposition 8.9: If Proposition 8.8 holds good, the best technical content of Enterprise 1 is

$$t_1^* = \frac{k(a - C)}{4\beta - k^2}$$

whereas that of Enterprise 2 is

$$t_1'^* = \frac{k(a-C)}{8\beta - k^2}$$

Proof: Obeying Proposition 8.8, Enterprise 1's and Enterprise 2's profits are differentiated with respect to the derivative of technical content, respectively. It follows that:

$$\frac{\partial \pi_1^*}{\partial t_1} = \frac{k(a-C+kt_1)}{4} - \beta t_1 = 0$$

Therefore

$$t_1^* = \frac{k(a-C)}{4\beta - k^2} \tag{8.21}$$

$$\frac{\partial \pi_2^*}{\partial t_1} = 2\frac{a-C+kt1}{4} * \frac{k}{4} = 0$$

$$t_1'^* = \frac{k(a-C)}{8\beta - k^2} \tag{8.22}$$

As can be seen from Formulas 8.21 and 8.22, in the dynamic game of complete information theory, the best technical content of enterprise innovation in advance is larger than that of the later enterprise innovation, which indicates that the higher the technical content of the innovator, the larger the barriers to entry, namely, it is harder for late comers to catch up. As a result, the innovators will be able to seize market opportunities and target users earlier, improve products and brand maturity faster, gain more competitive advantage compared to other manufacturers, and ultimately win larger profits.

2. Numerical analysis
   a. Impact of technical content on equilibrium profit of Enterprise 1

Assuming that $\alpha = 200$, $C = 10$, $k = 1$, $\beta = 100$ and technical content t_1 rises from 0.01 in steps of 0.02, by Proposition 8.8 the impact of technical content on equilibrium profit of Enterprise 1 can be obtained as shown in Figure 8.4.

   b. Impact of technical content on equilibrium profit of Enterprise 2

Assuming that $\alpha = 200$, $C = 10$, $k = 1$, $\beta = 100$ and technical content t_2 rises from 0.01 in steps of 0.02, by Proposition 8.8 the impact of technical content on equilibrium profit of Enterprise 1 can be obtained as shown in Figure 8.5.

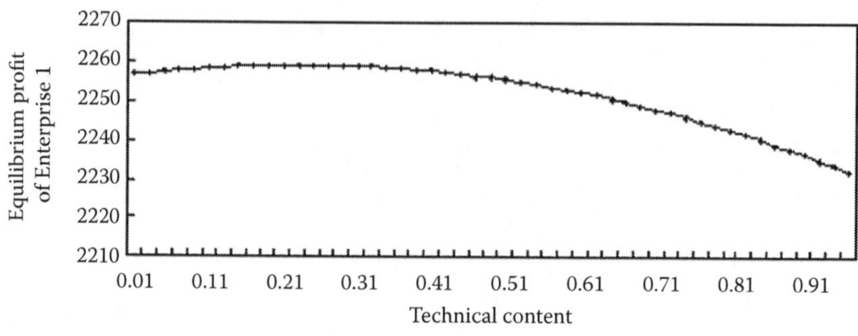

Figure 8.4 Impact of technical content on equilibrium profit of Enterprise 1.

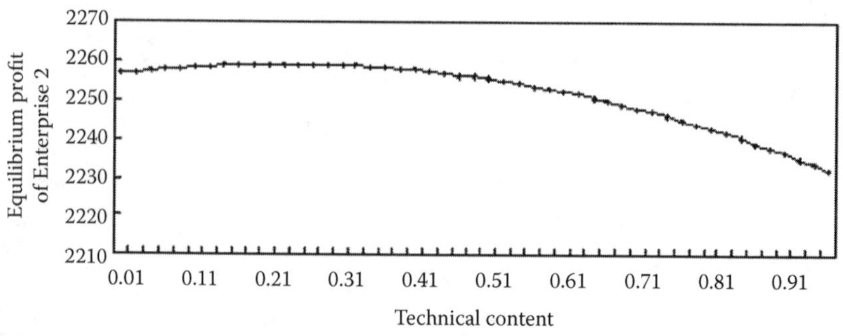

Figure 8.5 Technical content's impact on equilibrium profit of Enterprise 2.

As can be seen from Figures 8.4 and 8.5, in the dynamic game of complete information theory, there is a limit for enterprises to innovate, namely, the best technical content. However, in equal conditions, equilibrium profit due to the earlier innovation of Enterprise 1 is larger than that due to the later innovation of Enterprise 2; so is the case with the best technical content, which clearly shows that there is a large first-mover advantage in the process of innovation.

8.1.2.2 Game Model of Enterprises 1 and 2 with Different Technical Content after Innovation

1. Modeling and analysis

Assumption 8.5: Assume that technical content of Enterprise 1 after technology innovation is t_1, whereas that of Enterprise 2 after technology innovation is t_2.

Proposition 8.10: If the hypotheses of 8.1, 8.2, 8.3, 8.4, and 8.5 hold good, the equilibrium profit of Enterprise 1 is as follows:

$$\pi_1^* = \frac{(a - C + 2\mathrm{kt}_1 - kt_2)^2}{8} - \frac{1}{2}\beta t_1^2$$

The equilibrium profit of Enterprise 2 is as follows:

$$\pi_1^* = \frac{a - C + 2kt_1 - kt_2}{4} * \frac{a - C - 2kt_1 + 3kt_2}{4} - \frac{1}{2}\beta t_2^2$$

■

Proof: As can be seen from the Hypotheses of 8.1, 8.2, 8.3, 8.4, and 8.5, the profit of Enterprise 2 is as follows:

$$\begin{aligned}\pi_2 &= q_2(p(Q) - c_2) - \frac{1}{2}\beta t_2^2 \\ &= q_2(a - (q_1 + q_2) - C + kt_2) - \frac{1}{2}\beta t_2^2\end{aligned} \tag{8.23}$$

At first, equating $\frac{\partial \pi_2}{\partial q_2} = 0$ and considering q_1 to be decided, the best output of Enterprise 2 can be obtained as follows:

$$q_2^* = \frac{a - q_1 - C + kt_2}{2} \tag{8.24}$$

As Enterprise 1 can forecast for Enterprise 2 to choose q_2^*, its profit function is as follows:

$$\begin{aligned}\pi_1 &= q_1(a - (q_1 + q_2^*) - C + kt_1) - \frac{1}{2}\beta t_1^2 \\ &= q_1(a - q_1 - \frac{a - q_1 - C + kt_1}{2} - C + kt_1) - \frac{1}{2}\beta t_1^2\end{aligned} \tag{8.25}$$

$$\text{Let } \frac{\partial \pi_1}{\partial q_1} = \frac{a - 2q_1 - C + 2kt_1 - kt_2}{2} = 0$$

Therefore

$$q_1^* = \frac{a - C + 2kt_1 - kt_2}{2} \tag{8.26}$$

Using Formula 8.26 into Formula 8.24 the equilibrium output of Enterprise 2 is obtained as follows:

$$q_2^* = \frac{a - C - 2kt_1 + 3kt_2}{4} \tag{8.27}$$

As a result, within different technical content of innovation, the total output of the two enterprises is as follows:

$$Q = q_1^* + q_2^* = \frac{3a - 3C + 2kt_1 + kt_2}{4} \tag{8.28}$$

Using Formula 8.27 in Formula 8.25 the profit of Enterprise 1 is obtained as follows:

$$\pi_1^* = \frac{(a - C + 2kt_1 - kt_2)^2}{8} - \frac{1}{2}\beta t_1^2 \tag{8.29}$$

As can be seen from the equation above, larger t_2 means lower profits for Enterprise 1. When $t_2 = 0$, Enterprise 1 wins the greatest profit, which occurs when Enterprise 2 carries out no innovation. This is also obvious in real life.

Using Formulas 8.26 and 8.27 in Formula 8.23 the profit of Enterprise 2 is obtained as follows:

$$\pi_2^* = \frac{a - C + 2kt_1 - kt_2}{4} * \frac{a - C - 2kt_1 + 3kt_2}{4} - \frac{1}{2}\beta t_2^2 \tag{8.30}$$

❑

Let

$$\frac{\partial \pi_1^*}{\partial t_1} = \frac{k(a - C + 2kt_1 - kt_2)}{2} - \beta t_1 = 0$$

Therefore

$$t_1^* = \frac{k(a - C - kt_2)}{2\beta - 2k^2} \tag{8.31}$$

$$\frac{\partial^2 \pi_1^*}{\partial t_1^2} = k - \beta < 0$$

As can be seen from Formulas 8.30 and 8.31, there is a limit for Enterprise 1 to innovate, namely, the best technical content of innovation. Either lower or higher than the best technical content, the profit would reduce.

$$\frac{\partial t_1^*}{\partial t_2} = -\frac{k^2}{2\beta - 2k^2}$$

Because $\frac{\partial \pi_1^*}{\partial t_1} = \frac{k(a - C + 2kt_1 - kt_2)}{2} < 0$

Hence π_1^* is a decreasing function of t_2,

Add $t_2 \geq 0$

Therefore, when $t_2 = 0$ namely Enterprise 2 carries out no innovation, Enterprise 1 wins the greatest profit.

$$\frac{\partial \pi_2^*}{\partial t_2} = \frac{a - C + 2kt_1 - kt_2}{4} * \frac{3k}{4} - \frac{k}{4} * \frac{a - C - 2kt_1 + 3kt_2}{4} - \beta t_2 = 0$$

$$t_2^* = \frac{k(a - C + 4kt_1)}{8\beta + 3k^2} \qquad (8.32)$$

$$\frac{\partial \pi_2^*}{\partial t_1} = \frac{k}{2} * \frac{a - C - 2kt_1 + 3kt_2}{4} - \frac{k}{2} * \frac{a - C + 2kt_1 - kt_2}{4} = \frac{k^2(t_2 - t_1)}{2}$$

However, it is not the same with Enterprise 2. Unlike the relation of Enterprise 1's profit to Enterprise 2, that of Enterprise 2's profit to Enterprise 1's technical content is not so simple. By Formula 8.32, when $t_2 < t_1$, $\frac{\partial \pi_2^*}{\partial t_1} < 0$, that is, Enterprise 2's profit is a decreasing function of t_1, so it is better for t_1 to be smaller. As $t_1 \geq 0$, Enterprise 2 will win the greatest profit although neither Enterprise 1 nor Enterprise 2 carries out innovation. However, if $t_2 \geq t_1$, larger t_1 means larger t_2 and a greater profit for Enterprise 2. This shows that in the process of innovation, if the leading enterprise has higher technical content, it is easier for it to seize the market. Furthermore, if the enterprise that is catching up gets a much higher technical content, it is simple for it to win greater profit on the basis of the original enterprise.

1. Numerical analysis
   a. Impact of technical content of Enterprise 2 on equilibrium profit of Enterprise 1

Assuming that $\alpha = 200$, $C = 10$, $k = 1$, $\beta = 100$, $t_1 = 1$ and Enterprise 2's technical content t_2 rises from 0.01 in steps of 0.02 by Proposition 8.10 the impact of the technical content of Enterprise 2 on equilibrium profit of Enterprise 1 can be obtained as shown in Figure 8.6.

As can be seen from Figure 8.6, equilibrium profit of Enterprise 1 is inversely proportionate to the technical content of Enterprise 2. Although Enterprise 1 innovates earlier than Enterprise 2, larger technical content of Enterprise 2 means smaller profit for Enterprise 1.

   b. Impact of technical content of Enterprise 1 on equilibrium profit of Enterprise 2

Assuming that $\alpha = 200$, $C = 10$, $k = 1$, $\beta = 100$, $t_2 = 1$ and enterprise 1's technical content t_1 rises from 0.01 in steps of 0.02 by Proposition 8.10 that the impact of the technical content of Enterprise 1 on equilibrium profit of Enterprise 2 is obtained as shown in Figure 8.7.

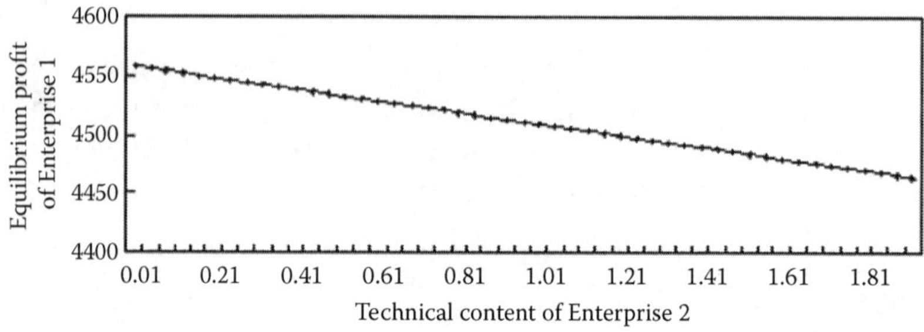

Figure 8.6 Technical content of Enterprise 2 and equilibrium profit of Enterprise 1.

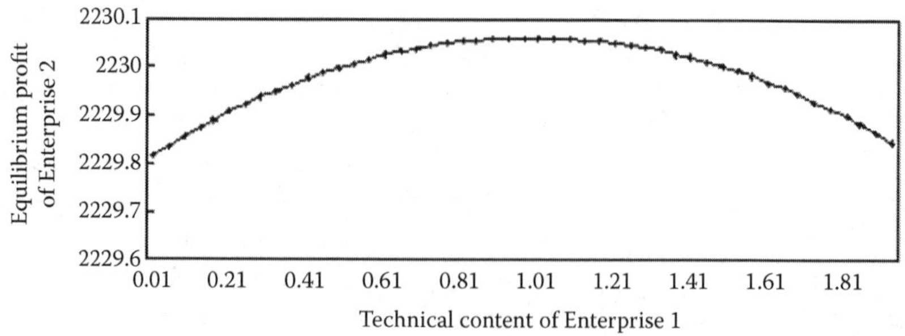

Figure 8.7 Technical content of Enterprise 1 and equilibrium profit of Enterprise 2.

As can be seen from Figure 8.7, the relation of Enterprise 2's profit to Enterprise 1's technical content is not inversely proportional, when Enterprise 2 has invariable ability of innovation, namely, its technical content is invariable even after innovation. When $t_2 > t_1$, larger t_1 indicates larger profit of Enterprise 2; When $t_2 < t_1$, Enterprise 2's profit is a decreasing function of t_1, so smaller t_1 is much better. It shows that in the process of innovation, if the leading enterprise has higher technical content, it is easier for it to seize the market. Furthermore, if the enterprise that is catching up gets a much higher technological content, it is simple for it to win greater profits on the basis of the original enterprise.

8.2 Game Model of Technology Transfer and Stable Strategy

8.2.1 Game Model and Stable Strategy of Technical Innovation of Fixed Payment Model

In true life, a scientific research unit has more technology innovations than an enterprise because of its larger number of talents; therefore because of the limitations of both scientific research and enterprise, any enterprise has to cooperate with a scientific research unit in technology innovation. Therefore, any enterprise frequently faces two choices: one is carrying out technology innovation by itself, the other is cooperating with a scientific research unit, namely, the enterprise pays the scientific research unit reciprocal money for innovation, and then makes use of those innovations to produce. If the enterprise and the scientific research unit carry out cooperative innovation, what is the game between them like?

8.2.1.1 Construction of Game Model

If the enterprise wants to carry out technology innovation, it could choose independent innovation or cooperation with a scientific research unit. Assuming that they choose their own strategies of action aiming at their own profit maximization with no moral hazard, a game model can be analyzed and solved.

8.2.1.2 Solution and Analysis of Game Model

1. Game model of scientific research unit innovating and Enterprise producing

---

Hypothesis 8.6: If the enterprise wants to use those innovations to produce, it has to pay scientific research unit reciprocal money, and reciprocal price is $p\left(p > \frac{1}{2}\beta t_1^2\right)$. Assuming that the profit of scientific research unit is a fixed proportion of the earnings, the profit rate is $r(0 < r < 1)$. ■

Proposition 8.11: If the Hypotheses of 8.2 and 8.6 hold good, the payment price is $p = \frac{\beta t_1^2}{2(1-r)}$, and the profit of the scientific research unit is $\pi_2 = rp = \frac{r\beta t_1^2}{2(1-r)}$. ■

Proof: By the Hypotheses of 8.2 and 8.6, the profit of scientific research unit is $\pi_2 = p - \frac{1}{2}\beta t_1^2 = rp$, the payment price is $p = \frac{\beta t_1^2}{2(1-r)}$ and $\pi_2 = rp = \frac{r\beta t_1^2}{2(1-r)}$.

Hence, proved.

As can be seen from Hypothesis 8.11, the payment price and the profit of scientific research unit are both quadratic functions of technical content, in other words, the higher the demand for innovation technical content, the higher price the enterprise needs to pay and the greater profit the scientific research unit gets.

Differentiating π_2 with respect to the technical content t_1 gives

$$\frac{\partial \pi_2}{\partial t_1} = \frac{r\beta t_1}{1-r}$$

As $\frac{\partial \pi_2}{\partial t_1} > 0$, the profit of scientific research unit is a increasing function of t1, which is obvious. Unless the enterprise buys technology from the scientific research unit, the latter cannot get any profit. What is more, the higher the technical content of innovation, the higher the price enterprise pays, and the greater the profit of the scientific research unit, assuming that the profit rate of the scientific research unit is fixed. As a result, if the ability of the scientific research unit allows, it is better for the technical content to be higher. ❑

Proposition 8.12: If the Hypotheses of 8.1, 8.3, and 8.11 hold good, the greatest profit of the enterprise is $\pi_1 = \left(\frac{a - C + kt_1}{2}\right)^2 - \frac{\beta t_1^2}{2(1-r)}$, and the best technical content is $t_1^* = \frac{k(a-C)}{(2\beta/(1-r)) - k^2}$. ■

Proof: By the Hypotheses of 8.2 and 8.6, it can be seen that the profit of the enterprise is as follows:

$$\pi_1 = q_1(a - q_1 - C + kt_1) - \frac{\beta t_1^2}{2(1-r)}$$

$$\frac{\partial \pi_1}{\partial q_1} = a - 2q_1 - C + kt_1 = 0 \tag{8.33}$$

The best output is as follows:

$$q_2^* = \frac{a - C + kt_1}{2}$$

The greatest profit is as follows:

$$\pi_1 = \frac{a - C + kt_1}{2}\left(a - \frac{a - C + kt_1}{2} - C + kt_1\right) - \frac{\beta t_1^2}{2(1-r)}$$

$$= \left(\frac{a - C + kt_1}{2}\right)^2 - \frac{\beta t_1^2}{2(1-r)} \tag{8.34}$$

Differentiating the profit with respect to the technical content and equating the derivative to 0:

$$\frac{\partial \pi_1}{\partial t_1} = 2\frac{a - C + kt_1}{2} \times \frac{k}{2} - \frac{\beta t_1}{1-r} = 0$$

The best technical content of Enterprise 1 in fixed payment model is:

$$t_1^* = \frac{k(a-C)(1-r)}{2\beta - (1-r)k^2} = \frac{k(a-C)}{\dfrac{2\beta}{1-r} - k^2} \tag{8.35}$$

Because $k > 0$, $a > C$, $1 - r > 0$, $t_1^* > 0$,
$2\beta - (1-r)k^2 > 0$

Therefore $\dfrac{\partial^2 \pi_1}{\partial t_1{}^2} = \dfrac{k^2}{2} - \dfrac{\beta}{1-r} = \dfrac{(1-r)k^2 - 2\beta}{2(1-r)} < 0$.

When the technical content is t_1^*, the function attains a maximum. So when the enterprise carries out independent innovation, the best technical content is as follows:

$$t_1^* = \frac{k(a-C)(1-r)}{2\beta - (1-r)k^2} = \frac{k(a-C)}{\dfrac{2\beta}{1-r} - k^2}$$

❑

As seen from Proposition 8.12, in the process of technical innovation, higher technical content does not mean larger profit. When the technical content is t_1^*, the innovation profit is greatest. In other words, innovation has its own limits and technology needs to advance steadily. A quick advance in technology requires too much investment, which will be more than the income. This is similar to some industries in real life. And t_1^* is inversely proportionate to the profit proportion r of the scientific research unit, that is, a higher r implies lower technical content.

8.2.1.3 Best Technical Content Analysis of Enterprise Production After Independent Technology Innovation

---

Proposition 8.13: If the Hypotheses of 8.1, 8.2, and 8.3 hold good, the best output of enterprise after independent technical innovation is $q_2 = \frac{a - C + kt_2}{2}$, and the greatest profit is $\pi'^* = \left(\frac{a - C + kt_2}{2}\right)^2 - \frac{1}{2}\beta t_2^2$, and, the best technical content is $t_2^* = \frac{k(a-C)}{2\beta - k^2}$. ■

Proof: As can be seen from the Hypotheses of 8.1, 8.2, and 8.3, the profit of the enterprise after independent technology innovation is as follows:

$$\pi' = q_2(p(q_2) - c_2) - \frac{1}{2}\beta t_2^2 \tag{8.36}$$

The best output of enterprise after technology innovation is as follows:

$$q_2^* = \frac{a - C + kt_2}{2} \tag{8.37}$$

Using Formula 8.37 in Formula 8.36, the largest profit of enterprise after independent technology innovation is as follows:

$$\pi'^* = \left(\frac{a - C + kt_2}{2}\right)^2 - \frac{1}{2}\beta t_2^2$$

$$\frac{\partial \pi'^*}{\partial t_2} = 2\frac{a - C + kt_2}{2} * \frac{k}{2} - \beta t_2 = 0 \tag{8.38}$$

Therefore

$$t_2^* = \frac{k(a-C)}{2\beta - k^2} \tag{8.39}$$

❑

As $1 - r < 1$, we can obtain $t_1^* < t_2^*$ from Propositions 8.12 and 8.13, which shows that a higher technical content of the enterprise after independent technology innovation is better for technology advancement and resource transformations. If the technical contents are equal, the profit is larger when the enterprise carries out independent technology innovation.

8.2.1.4 Numerical Analysis

1. Impact of production technical content on the best profit of enterprise

We assume that $\alpha = 200$, $C = 10$, $k = 1$, $\beta = 100$, $r = 0.2$ and technical content rises from 0.012 in steps of 0.02. By Proposition 8.12 the impact of product technical content on the best profit can be obtained as shown in Figure 8.8.

As can be seen from Figure 8.8, the best profit of the enterprise improves along with the increase of technical content at first, but declines after a period of time, as the foregoing conclusion shows.

2. Impact of profit rate r on technical content t

Assuming that $\alpha = 200$, $C = 10$, $k = 1$, $\beta = 100$ and the profit rate rises from 0.01 in steps of 0.01, by Proposition 8.12 the impact of the profit rate on technical content can be obtained as shown in Figure 8.9.

As can be seen from Figure 8.9, the technical content is inversely proportionate to the profit rate of the scientific research unit, that is, a larger profit rate indicates lower technical content. As in the case of funds, a larger profit rate means lesser funds for research, and naturally leads to lower technical content of innovation.

3. Impact of marginal technical content's input parameter β on the best technical content

Assuming that $\alpha = 200$, $C = 10$, $k = 1$, $r = 0.2$ and the marginal technical content's input parameter β rises from 50 in steps of 5, by Proposition 8.12 the impact of

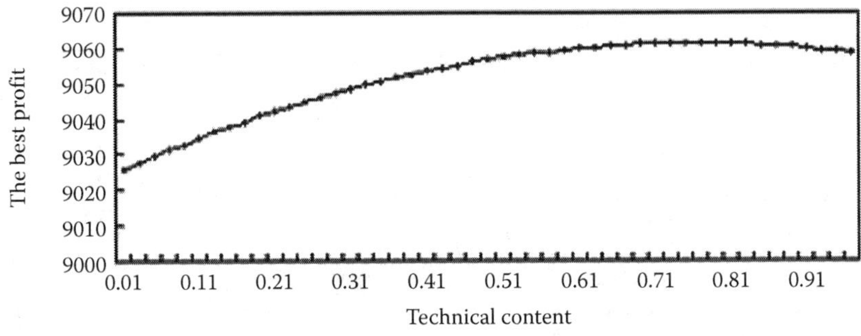

Figure 8.8 Impact of technical content on the best profit.

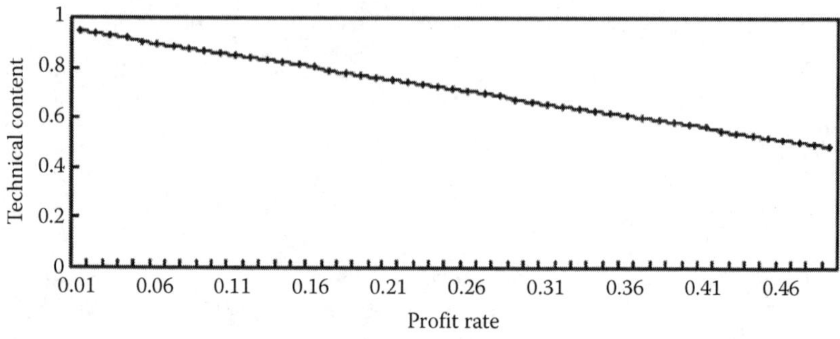

Figure 8.9 Impact of the profit rate on technical content.

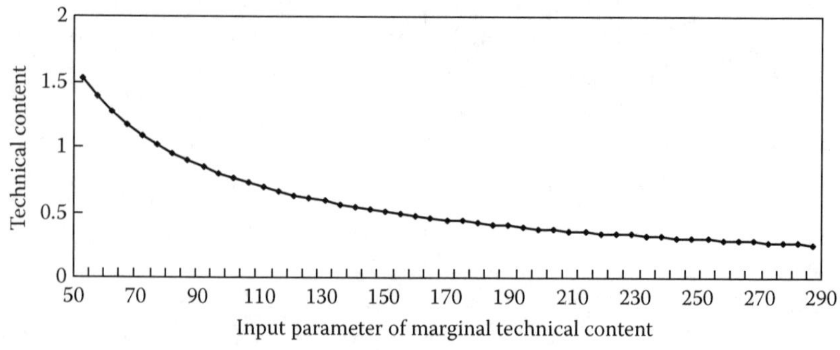

Figure 8.10 Impact of input parameter of marginal technical content on technical content.

the marginal technical content's input parameter β on the technical content can be obtained as shown in Figure 8.10.

As can be seen from Figure 8.10, the best technical content is inversely proportionate to the marginal technical content's input parameter β, that is, a larger β indicates lower technical content. This shows that if an enterprise is a large industry and achieves a certain degree of technical content, its technology will be difficult for other enterprises to imitate. As a result, this enterprise will monopolize the marker for a long time to win more profit.

4. Impact of the cost decreased coefficient of unit product technical content on the best technical content

Assuming that $\alpha = 200$, $C = 10$, $\beta = 100$, $r = 0.2$, unit product technical content's cost decreased coefficient k rises from 0.01 in steps of 0.04, by Proposition 8.12 that the impact of the unit product technical content's cost decreased coefficient on the best technical content can be obtained as shown in Figure 8.11.

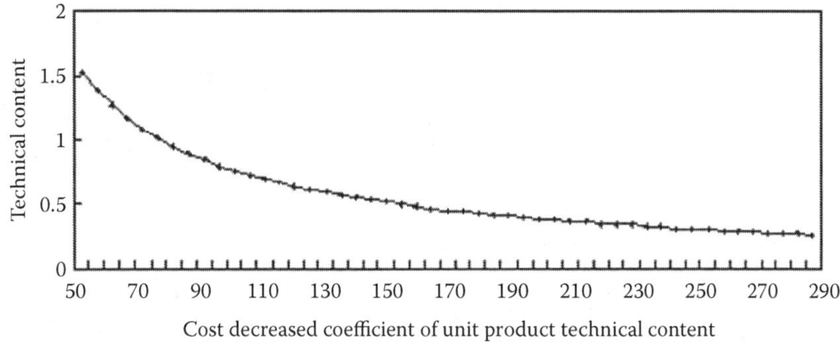

Figure 8.11 Impact of the cost decreased coefficient of unit product technical content on technical content.

As can be seen from Figure 8.11, the best technical content is proportional to the unit product technical content's cost decreased coefficient k. If others conditions remain unchanged, a larger k means a larger technical content can be achieved. This is understandable, as after innovation, the lower the cost, the more the profit of the enterprise, and it is more beneficial to increase investment to get a higher level of the best technical content.

8.2.2 Game Model and Stable Strategy of Technology Innovation while Paying according to a Certain Percentage of Total Profit

8.2.2.1 Modeling and Analysis

In the model above, the sales risk of product was not considered and only the market-clearing price was taken into account. In reality, the product may remain unsold. Furthermore, the profit from sales has a time lag. As a result, in the game model, the profits of an enterprise appear to be larger, but in practice the profit of the scientific research unit is more reliable, as the enterprise would have to pay funds for research before innovation. Therefore, to share the risk, the enterprise will sometimes take a different form of payment, that is research money paid to the scientific research unit will not be a fixed amount. But relying on the output of the product, the greater the sales amount of the product, the more the money that will be paid to the scientific research unit.

Hypothesis 8.7: The technical content of research-based innovative product is t, and the enterprise pays scientific research unit a certain percentage of the sales amount. The royalty rate is μ. ■

Proposition 8.14: If the Hypotheses of 8.1, 8.2, 8.3, and 8.7 hold good, the equilibrium output is $q_1^* = \frac{a-C+kt}{2}$, and the profits of the enterprise and the scientific

$$\pi_1^* = (1-\mu)\left(\frac{a-C+kt}{2}\right)^2$$

research unit are separately $\pi_2^* = \mu\left(\frac{a-C+kt}{2}\right)^2 - \frac{1}{2}\beta t^2$, when, the best technical content of the scientific research unit is $t^* = \frac{\mu k(a-C)}{2\beta - \mu k^2}$. ■

Proof: As can be seen from the Hypotheses of 8.1, 8.2, 8.3, and 8.7, the profit of Enterprise 1 is as follows:

$$\pi_1 = (1-\mu)q_1(a - q_1 - C + kt) \tag{8.40}$$

The profit of the scientific research unit is as follows:

$$\pi_2 = \mu q_1(a - q_1 - C + kt) - \frac{1}{2}\beta t^2 \tag{8.41}$$

By Formula 8.40

$$\frac{\partial \pi_1}{\partial q_1} = (1-\mu)(a - 2q_1 - C + kt) = 0$$

Hence

$$q_1^* = \frac{a-C+kt}{2} \tag{8.42}$$

Therefore

$$\begin{aligned}\pi_1^* &= (1-\mu)\frac{a-C+kt}{2}\left(a - \frac{a-C+kt}{2} - C + kt\right) \\ &= (1-\mu)\left(\frac{a-C+kt}{2}\right)^2\end{aligned} \tag{8.43}$$

$$\pi_2^* = \mu\frac{a-C+kt}{2}\left(a-\frac{a-C+kt}{2}-C+kt\right)-\frac{1}{2}\beta t^2$$

$$=\mu\left(\frac{a-C+kt}{2}\right)^2-\frac{1}{2}\beta t^2 \tag{8.44}$$

For $\frac{\partial\pi_1^*}{\partial t}=2(1-\mu)\frac{a-C+kt}{2}*\frac{k}{2}>0$

So π_1 is a increasing function of technical content t, that is, it is better for t to get larger.

$$\frac{\partial\pi_2^*}{\partial t}=2\mu\frac{a-C+kt}{2}*\frac{k}{2}-\beta t=0$$

It can be deduced that:

$$t^*=\frac{\mu k(a-C)}{2\beta-\mu k^2} \tag{8.45}$$

❑

As can be seen from Proposition 8.14, the technical content t is a increasing function of μ, which shows that a larger royalty rate μ indicates a larger technical content t. It can be deduced that, when the royalty rate for the scientific research unit is larger, the funds for research are larger, and the technical content is certainly higher. However, the royalty rate cannot be too large, as Enterprise 1 needs to hold on to its normal profit. So the royalty rate should be decided by specific conditions, in close relation with reality. If only a small part of the scientific research can be exploited, the royalty rate for the scientific research unit will be high; if the item of technology is not difficult to exploit, the royalty rate will be lower; the royalty rate, therefore, is determined by the supply and demand in the market economy.

$$\frac{\partial^2\pi_2^*}{\partial t^2}=\frac{\mu k^2}{2}-\beta \tag{8.46}$$

For $t^*=\frac{\mu k(a-C)}{2\beta-\mu k^2}>0$

Therefore $\mu k^2<2\beta$, namely $\partial^2\pi_2/\partial t_2>0$. As a result, the greatest profit of the scientific research unit is when the technical content is t^* and t^* is the best technical content of the scientific research unit. In other words, when the input parameter of marginal technical content of the scientific research unit is larger than the product of cost decreased coefficient's second power and the royalty rate, it is not always better to improve innovation and the best technical content. If the input parameter of marginal technical content in a certain item of technology is very large or the royalty rate is too small, the profit of the scientific research unit will reach its maximum at a certain technical content, that is, at either lower or higher than this certain value, the profit will decrease.

8.2.2.2 Numerical Analysis

1. Impact of technical content on the profit of Enterprise 1 when the royalty rate is fixed

Assuming that $\alpha = 200$, $C = 10$, $k = 1$, $\beta = 100$, $\mu = 0.3$ and technical content rises from 0.01 in steps of 0.02, by Proposition 8.14 the impact of the technical content on the profit of Enterprise 1 can be obtained as shown in Figure 8.12.

As can be seen from Figure 8.12, when the enterprise pays the scientific research unit a certain proportion of the total sales amount in the process of technology transfer there is no best technical content of innovation, unlike in enterprises carrying out independent innovation. The profit of the enterprise is proportional to the technical content, that is, it is better for the technical content to increase.

2. Impact of technical content on the profit of scientific research unit when the royalty rate is fixed

Assuming that $\alpha = 200$, $C = 10$, $k = 1$, $\beta = 100$, $\mu = 0.3$ and technical content rises from 0.01 in steps of 0.02, by Proposition 8.14 the impact of the technical content on the profit of the scientific research unit can be obtained as shown in Figure 8.13.

As can be seen from Figure 8.13, for the scientific research unit, it is not always better to improve innovation and the best technical content.

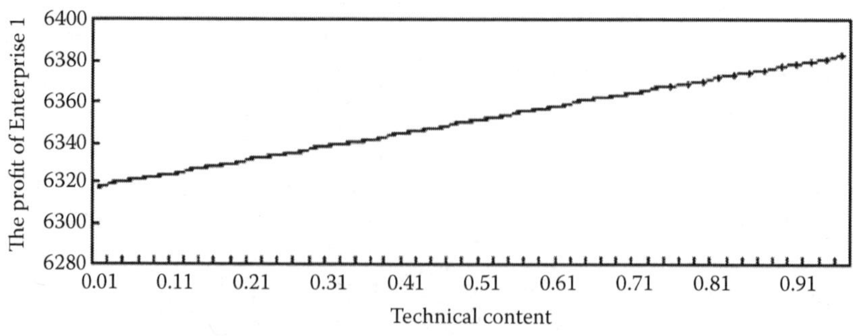

Figure 8.12 Impact of technical content on the profit of Enterprise 1.

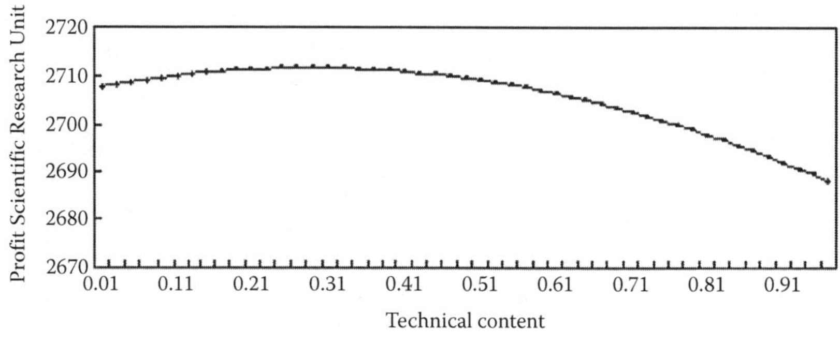

Figure 8.13 Impact of technical content on the profit of scientific research unit.

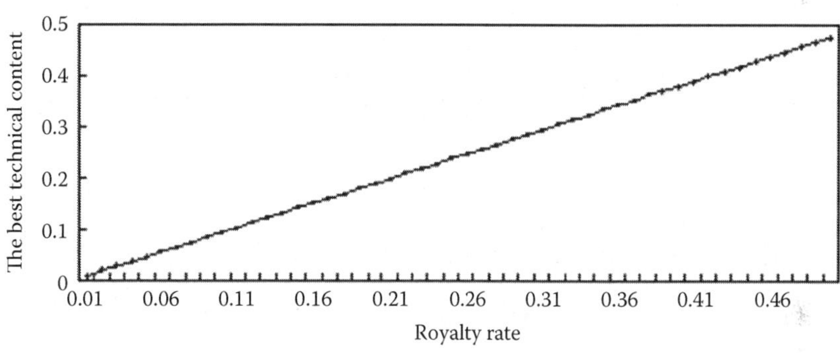

Figure 8.14 Impact of royalty rate on the best technical content.

3. Impact of Royalty Rate on the best technical content

Assuming that $\alpha = 200$, $C = 10$, $k = 1$, $\beta = 100$, and technical content rises from 0.01 in steps of 0.01, by Figure 8.14 the impact of the royalty rate on technical content can be obtained as shown in Figure 8.14.

As can be seen from Figure 8.14, the best technical content is proportional to the royalty rate. In other words, a higher royalty rate for the scientific research unit means a larger technical content of innovation.

8.3 Game Model of Technical Diffusion and Its Stable Strategy Analysis

The diffusion of technical innovation can be seen as the process that enterprises with potential requirements of innovative technologies adopt, motivated by the abduction of excess monopoly profits. When an innovative technology has not been completely diffused, but, at the same time is already in use by some enterprises, then a short-term equilibrium will be formed under monopolistic competition. In this condition, these enterprises will get excess monopoly profits, because the price is higher than the average cost in that period.

In practice, more and more new enterprises will be attracted to enter this field, leading to the final price becoming equal to the average cost. Then the excess monopoly profit will no longer exist and no more enterprises will enter the field. Then long-term equilibrium, called the "fully diffused status," will be reached, and diffusion of technology will stop.

Therefore, the process of proliferation can be divided into three stages.

8.3.1 Single Monopoly of Innovator

The first stage: single monopoly of the innovator. Assuming that Enterprise 1 is the only one who enters the market with a particular innovative technology, because the followers cannot produce the same products in a short period, at that point of time, Enterprise 1 will enjoy optimal production with the optimal price, and monopolize the profits in the market.

Hypothesis 8.8: It is assumed that the demand of the market is Q, and the technical content of innovated products by Enterprise 1 is t_1. ■

Proposition 8.15: If Hypotheses 8.1, 8.2, 8.3, and 8.8 are tenable, the optimal quantity, profit, and technical content are respectively $Q^* = \frac{a - C + kt_1}{2}$, $\pi_1^* = \left(\frac{a - C + kt_1}{2}\right)^2 - \frac{1}{2}\beta t_1^2$, and $t_1^* = \frac{k(a - C)}{2\beta - k^2}$. ■

Proof: From Hypotheses 8.1, 8.2, 8.3, and 8.8, the profit of Enterprise 1 is

$$\pi_1 = Q(p(Q) - c_1) - \frac{1}{2}\beta t_1^2 \tag{8.47}$$

Taking the first derivative of Formula 8.47 with respect to Q and equating the derivative to 0,

$$\frac{\partial \pi_1}{\partial Q} = a - 2Q - C + kt_1 = 0$$

The optimal quantity of technical innovation is as follows:

$$Q^* = \frac{a - C + kt_1}{2} \tag{8.48}$$

The optimal profit of technical innovation is as follows:

$$\pi_1^* = \left(\frac{a - C + kt_1}{2}\right)^2 - \frac{1}{2}\beta t_1^2 \tag{8.49}$$

$$\frac{\partial \pi_1^*}{\partial t_1} = 2\frac{a - C + kt_1}{2} * \frac{k}{2} - \beta t_1 = 0$$

Then,

$$t_1^* = \frac{k(a - C)}{2\beta - k^2} \qquad (\text{because } t_1 > 0, a > 0, 2\beta - k^2 > 0)$$

So

$$\frac{\partial^2 \pi_1^*}{\partial t_1{}^2} = \frac{k^2}{2} - \beta < 0$$

❑

From Proposition 8.15, we can see that, when $t < t_1^*$, the profit of Enterprise 1 increases with the increase of t, and it increases rapidly in the beginning, then increases slowly, and so Enterprise 1 can get its optimal profit at the point $t_1^* = \frac{a - C}{2\beta - k^2}$. When t is greater than t_1^*, the profit decreases with the increase of t, and it decreases slowly in the beginning, then decreases rapidly.

8.3.2 Another Company Entering the Market

As an effect of diffusion, in the process of industrialization of new technologies, many enterprises will dedicate time and money to research and development on the products with monopoly profits, and so the monopoly of Enterprise 1 will be temporary. There are bound to be other companies that enter the market with their developed products. The duopoly competition will come into play after the second company enters the market, though it may not always be a success in innovating and then the technical content of the innovated product of the second company will be $t_2 (t_2 \leq t_1)$.

In this case of Cournor static game with incomplete information, Enterprise 2 is aware of the technical content of Enterprise 1 and its profits. However, Enterprise 1 does not know the degree of imitation in the new products of Enterprise 2 or its technical level.

Hypothesis 8.9: There are two possible results to Enterprise 2: one is that it succeeds in imitating. In this situation, the probability is x and the technical level is t_1.The other is that the imitation is not successful, as a result, the probability is $1 - x$, and the technical level is t_2. The outputs are decided by both enterprises. ■

Proposition 8.16: If Hypotheses 8.1, 8.2, 8.3, and 8.9 are tenable, the profits of Enterprises 1 and 2 in the secondary stage are respectively

$$\pi_1^* = \left(\frac{a - C + 2kt_1 - kt_2 - kx(t_1 - t_2)}{3}\right)^2$$

$$\pi_2 = x\left(\frac{2a - 2C + kt_1 + kt_2 + kx(t_1 - t_2)}{6}\right)^2 + (1-x)\left(\frac{2a - 2C - 2kt_1 + 4kt_2 + kx(t_1 - t_2)}{6}\right)^2 +$$

$$-\frac{1}{2}x\beta t_1^2 - \frac{1}{2}(1-x)\beta t_2^2$$

■

Proof: From Hypotheses 8.1, 8.2, 8.3, and 8.9, we can get the profits of Enterprise 2, as follows:

$$\pi_2 = q_2^*(a - (q_1 + q_2) - C + kt_i) - \frac{1}{2}x\beta t_1^2 - \frac{1}{2}\beta(1-x)t_2^2 \tag{8.50}$$

Let: $\dfrac{\partial \pi_2}{\partial q_2} = a - q_1 - 2q_2 - C + kt_i = 0$

The reflected function of Enterprise 2's output relative to Enterprise 1's is

$$q_2^* = \frac{a - q_1 - C + kt_i}{2} \tag{8.51}$$

If Enterprise 2 imitates successfully, the output is q_2^H, otherwise, the output is q_2^L.

$$q_2^H = \frac{a - q_1 - C + kt_1}{2} \tag{8.52}$$

$$q_2^L = \frac{a - q_1 - C + kt_2}{2} \tag{8.53}$$

Because Enterprise 1 does not know the technical content of Enterprise 2, it should take the expected effectiveness into consideration, when making the decision.

Then

$$E\pi_1 = E[a - q_1 - q_2 - C + kt_1) * q_1]$$

$$= x * (a - q_1 - q_2^H - C + kt_1) * q_1 + (1 - x) * (a - q_1 - q_2^L - C + kt_1) * q_1$$

$$\frac{\partial E(\pi_1)}{\partial q_1} = x.(a - q_1 - q_2^H - C + kt_1) - X.q_1 + (1 - x).(a - q_1 - q_{2L} - C + kt_1) - (1 - x).q_1$$

$$= a - q_1 - q_2^L - C + kt_1 - xq_2^H + xq_2^L - q_1$$

$$= a - 2q_1 - C + kt_1 - (1 - x).q_2^L - x.q_2^H = 0$$

The deflected function of q_1 is as follows:

$$q_1^* = \frac{a - C + kt_1 - (1 - x)q_2^L - xq_2^H}{2} \tag{8.54}$$

Combining the functions of Formulas 8.52, 8.53, and 8.54, we can get the solution:

$$q_1^* = \frac{a - C + 2kt_1 - kt_2 - kx(t_1 - t_2)}{3}$$

$$= \frac{a - C + 2kt_1 - k(t_2 + xt_1 - xt_2)}{3} \tag{8.55}$$

Let $F(x) = t_2 + xt_1 - xt_2$, $F'(x) = t_1 - t_2$, because $t_1 > t_2$, $F(x)$ is an increasing function of x.

From Formula 8.55, there is negative correlation between the optimal output q_1^* of Enterprise 1 and the probability X of Enterprise 2 to successfully innovate. That is the more X increases the less q_1^* will be.

When other conditions remain unchanged, the output of Enterprise 1 is related to the technical level t_2 of imitation of Enterprise 2, which means the closer t_2 reaches the level of Enterprise 1's, the less q_1^* will be.

Substituting Formula 8.55 in Formula 8.52:

$$q_2^H = \frac{2a - 2C + kt_1 + kt_2 + kx(t_1 - t_2)}{6}$$

$$= \frac{2a - 2C + kt_1 + k(t_2 + xt_1 - xt_2)}{6} \tag{8.56}$$

There is proportional relationship between q_2^H and x, that is the higher the probability of successful innovation, the higher the output q_2^H of Enterprise 2 will be. Substituting Formula 8.55 in Formula 8.53:

$$q_2^L = \frac{2a - 2C - 2kt_1 + 4kt_2 + kx(t_1 - t_2)}{6} \tag{8.57}$$

There is a negative correlation between q_2^H and x, which implies that the higher the probability of successful innovation, the lower the output q_2^L of Enterprise 2 will be.

The following are the profits of Enterprise 2

$$\pi_2 = q_2^*(a - (q_1 + q_2) - C + kt_1) - \frac{1}{2}x\beta t_1^2 - \frac{1}{2}\beta(1 - x)t_2^2$$

$$= xq_2^H(a - (q_1 + q_2^H) - C + kt_1) - \frac{1}{2}x\beta t_1^2 + (1 - x)q_2^L(a - (q_1 + q_2^L) - C + kt_2) - \frac{1}{2}\beta t_2^2$$

$$= x\left(\frac{2a - 2C + kt_1 + kt_2 + kx(t_1 - t_2)}{6}\right)^2 + (1 - x)\left(\frac{2a - 2C - 2kt_1 + 4kt_2 + kx(t_1 - t_2)}{6}\right)^2$$

$$-\frac{1}{2}x\beta t_1^2 - \frac{1}{2}(1 - x)\beta t_2^2$$

$$= x(q_2^H)^2 + (1 - x)(q_2^H)^2 - \frac{1}{2}x\beta t_1^2 - \frac{1}{2}(1 - x)\beta t_2^2$$

The following are the equilibrium profits of Enterprise 1 in the secondary stage:

$$\pi_1^* = x * (a - q_1 - q_2^H - C + kt_1)q_1 + (1 - x)(a - q_1 - q_2^L - C + kt_1)q_1$$

$$= x\frac{2a - 2C + kt_1 + kt_2 + kx(t_1 - t_2)}{6}q_1 + (1 - x)\frac{2a - 2C - 2kt_1 + 4kt_2 + kx(t_1 - t_2)}{6}q_1$$

$$= \left(\frac{a - C + 2kt_1 - kt_2 - kx(t_1 - t_2)}{3}\right)^2$$

❑

Moreover:

$$\frac{\partial \pi_1^*}{\partial t_1} = 2\frac{a - C + 2kt_1 - kt_2 - kx(t_1 - t_2)}{3} * \frac{2k - kx}{3}$$
$$= \frac{2k(2-x)[a - C + kt_1(2-x) - kt_2(1-x)]}{9}$$

Assuming that $t_2 \leq t_1$, $\frac{\partial \pi_1^*}{\partial t_1} > 0$, that is, in the event of another enterprise entering the market, the profits of Enterprise 1 in the secondary stage is proportional to the innovative technical content t_1 in the first monopoly stage, which means the more the technical content of innovation in the first stage, the more the profits of Enterprise 1 will be in the secondary stage.

$$\frac{\partial \pi_2}{\partial t_2} = \frac{kx(1-x)[2a - 2C + kt_1 + kt_2 + kx(t_1 - t_2)]}{3}$$
$$+ \frac{k(1-x)(4-x)[2a - 2C - 2kt_1 + 4kt_2 + kx(t_1 - t_2)]}{3} - (1-x)\beta t_2$$
$$= \frac{k(1-x)(8a - 8C - 8kt_1 + 16kt_2 + 7kx(t_1 - t_2)}{3} - (1-x)\beta t_2$$

$$\frac{\partial \pi_2}{\partial t_2} = 0$$

From the above results it is clear that:

1. If $x = 1$, the technical content of Enterprise 2 is the same as Enterprise 1. At this time, $\frac{\partial \pi_2}{\partial t_2}$ is equal to 0. Because the technical content of Enterprise 1 is fixed, Enterprise 2 could reach its optimal profits only when it obtains the same technical content as Enterprise 1.
2. If $x \neq 1$, then

$$t_2^* = \frac{8a - 8C - 8kt_1 + 7kxt_1}{16k^2 - 7k^2x - 3\beta} \tag{8.58}$$

From the above equation, it is clear that the technical content of Enterprise 2 is negatively correlated with that of Enterprise 1.

8.3.3 N Enterprises Entering the Market

With further diffusion of technology, more related products are developed, and more enterprises enter the market. Then, a competitive climate of N competitors is created, through mutual game play. In this situation, the technology has been widely used, so the companies are close to each other in the aspect of technical level, which is in accordance with the Cournot multioligopoly competitive model. Simplifying the calculation, the technical content of each enterprise is t_1, and any enterprise should pay the cost so as to produce the products with technical level t_1.

Hypothesis 8.10: It is assumed that in the competition of N enterprises, the technical content of each is t_1, Besides, to each enterprise, marginal parameters of learning is λ, and learning cost is $P=\frac{1}{2}\lambda t_i^2 (0<\lambda<\beta,\quad 0<i\leq N)$. ■

Proposition 8.17: If Hypotheses 8.1, 8.2, 8.3, and 8.10 are tenable, the optimal technical content of the Cournot competitive model is as follows:

$$t_1^*=\frac{2k(a-C)}{(N+1)^2\lambda-2k^2}$$ ■

Proof: From Hypotheses 8.1, 8.2, 8.3, and 8.10, the profits of Enterprise i, are as follows:

$$\begin{aligned}\pi_i &= q_i^*\, p(Q)-q_i^*\, c_i-\frac{1}{2}\lambda t_i^2\\ &= q_i(p(Q)-c_1)-\frac{1}{2}\lambda t_1^2\\ &= q_i\left(a-\sum_{j=1}^{N}q_j-C+kt_1\right)-\frac{1}{2}\lambda t_1^2\end{aligned} \tag{8.59}$$

Then

$$q_i=\frac{a-C+kt_1}{N+1}\quad (0<i\leq N) \tag{8.60}$$

$$\pi_i^*=\left(\frac{a-C+kt_1}{N+1}\right)^2-\frac{1}{2}\lambda t_1^2\quad (0<i\leq N)$$

Taking the first derivative with respect to t_1 in the above equation, and equating the derivative to 0,

$$\frac{\partial \pi_i}{\partial t_1} = 2\frac{a - C + kt_1}{3} * \frac{k}{N+1} - \lambda t_1 = 0$$

Then:

$$t_1^* = \frac{2k(a-C)}{(N+1)^2\lambda - 2k^2} \tag{8.61}$$

❑

It is clear from Proposition 8.17 that, when other conditions remain unchanged, t_1 is negatively correlated with N. Then, the more the technology content of technology innovation the more the enterprises that will participate in the technology spillover, and vice versa. Subsequently, if technology content is steady, the more the enterprises that participate the less the profit will be.

Because of the restriction of the assumptions, in the model, the output of each enterprise in the total competition will be the same. However, in reality, as the difference in the scale of the enterprise and the starting time, the output of different enterprises may differ. On the contrary, as the technology has been widely used, the price of the product will remain nearly the same. With the decrease in profits, large enterprises, which entered the market early, may gradually reduce their output, and move to other products with higher technical content. But to small enterprises, where the cost of imitation is small, the profits are acceptable. The assumption, treating the products of different enterprises as the same, can serve as a general rule.

8.3.4 Numerical Analysis

8.3.4.1 Comparison of Profits in Different Stages of Enterprise 1

It is assumed that $\alpha = 200$, $C = 10$, $k = 1$, $\beta = 100$, $\lambda = 1$, $N = 20$, t_1, t_2, and x, respectively rises from 0.5, 0, and 0, in steps of 0.02, 0.03, and 0.02 respectively. From Proposition 5.1, the profit of Enterprise 1 in the first stage can be obtained, as shown in Figure 8.15. Furthermore, from Proposition 8.16, the profit of Enterprise 1 in the second stage can be calculated, as shown in Figure 8.16. From Proposition 5.1, the profit of Enterprise 1 in the third stage can be calculated, as shown in Figure 8.17.

From Figures 8.15 through 8.17, the following conclusions can be drawn: first, Enterprise 1 gets its highest profits in the first stage and its least profits in the third stage, which means the monopoly enterprise has the advantages of monopoly. Second, in the first and the third stage, Enterprise 1 has optimal technical content of innovation. However, in the second stage, the higher the technical content it has

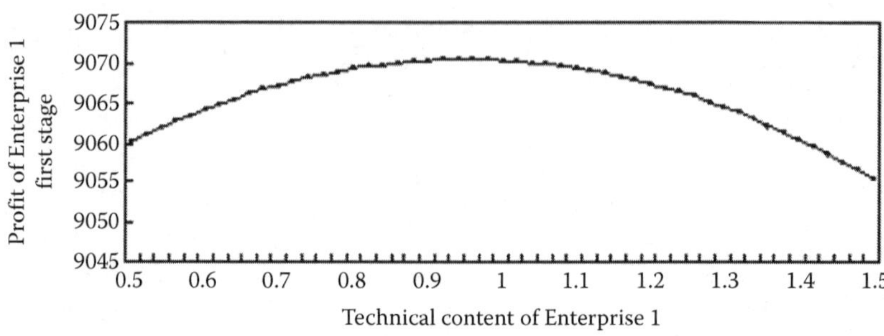

Figure 8.15 Profit of Enterprise 1 in the first stage.

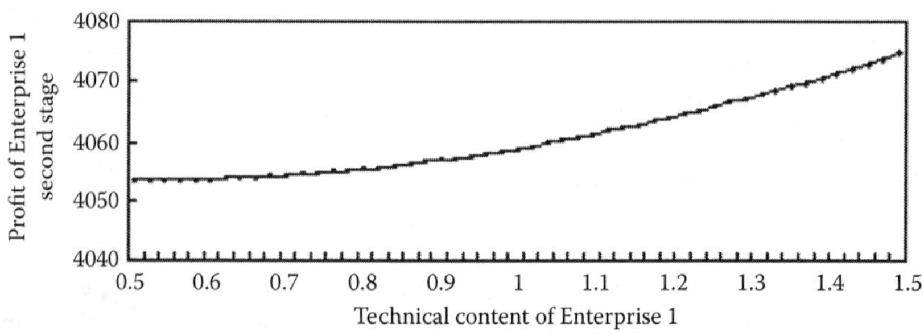

Figure 8.16 Profit of Enterprise 1 in the second stage.

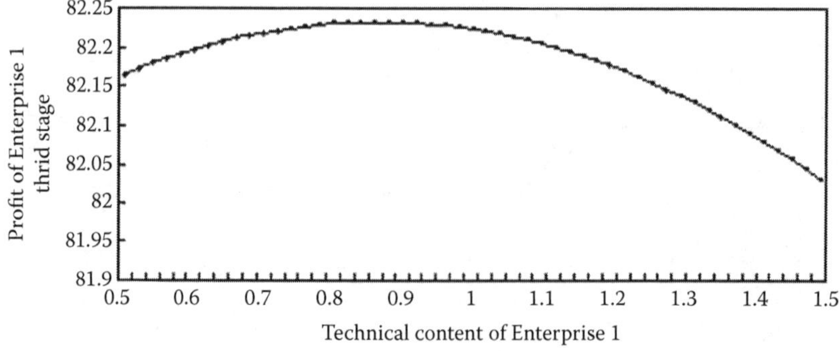

Figure 8.17 Profit of Enterprise 1 in the third stage.

the better, as the higher the technical content it has the harder it is for Enterprise 2 to compete with it. Finally, in the third stage, the profit of Enterprise 1 is low. So to get further advantages in the keen competition of the market, Enterprise 1 should venture into the next round of innovation.

8.3.4.2 Analysis on Gross Profit

Assume that $\alpha = 200$, $C = 10$, $k = 1$, $\beta = 100$, $\lambda = 1$, $N = 20$, t_1, t_2, and x respectively rises from 0.5, 0, and 0, in steps of 0.02, 0.03, and 0.02 respectively. From Proposition 8.16, gross profit in the second stage, can be obtained as shown in Figure 8.18. Furthermore, from Proposition 8.17, gross profit in the third stage can be obtained, as shown in Figure 8.19.

In the first stage, only one enterprise exists, so gross profit is equal to the profit of Enterprise 1. From Figure 8.15, Figures 8.18 and 8.19, the following two points are clear:

First, the optimal technical content of innovation is required for gross profit in every stage, but the requirements differ. Gross profit in the first stage demands the

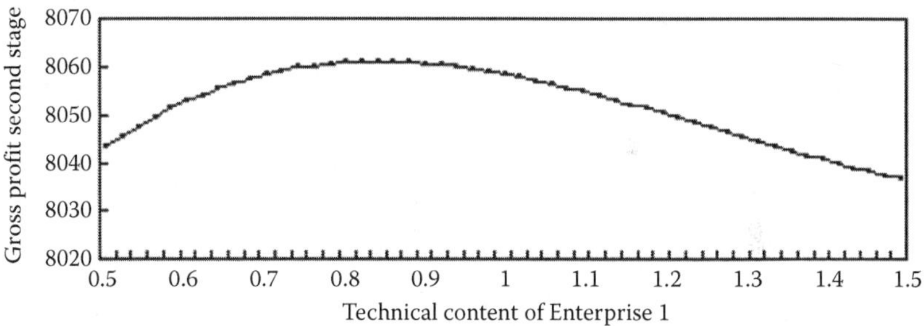

Figure 8.18 Gross profits in the second stage.

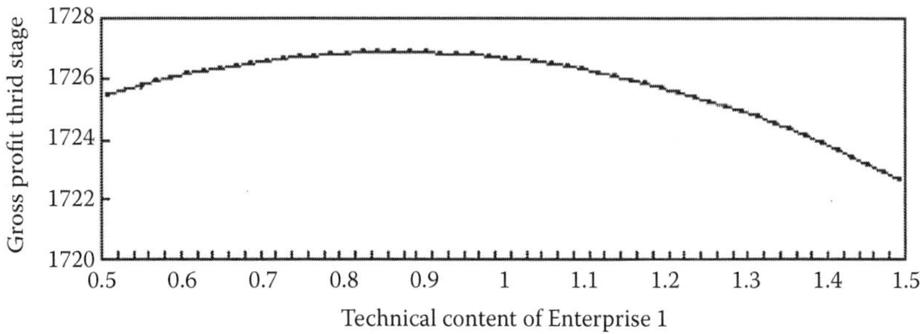

Figure 8.19 Gross profits in the third stage.

most, followed by that in the second stage, and the minimum is required in the third stage. Second, after the innovation of all the enterprises, the gross profit in the first stage is higher than that in the second stage, which is in turn higher than that in the third stage. That is to say, with the diffusion of technology, the profit of enterprises has decreased; this has been transformed into consumer surplus, and social welfare improves.

8.4 Game Model of Technical Spillover and Its Stable Strategy Analysis

In real life, there are only a few products that are exactly the same as the others. Even though the products have the same technical content, the brand, the technology, and some other aspects may differ. In addition, the spillover of technology is inevitable. This chapter focuses mainly on the Game model of technical innovation and its strategy analysis, under the condition that the spillover of innovation exists, and that the products are alternative. For the convenience of the discussion, technical innovation of duopoly enterprises are taken into consideration.

8.4.1 Construction and Analysis of the Model

Hypothesis 8.11: The products are different and substitutable. The alternative coefficient is $\theta(0 \le \theta \le 1)$. At the same time, the spillover of innovation exists, and the spillover coefficient is $\delta(0 \le \delta < 1)$. ■

Proposition 8.18: If Hypotheses 8.1, 8.2, 8.3, and 8.11 are tenable, the profits of Enterprise i will be as follows.

$$\pi_i = (a - q_i - \theta q_j - C + kt_i + \delta kt_j)q_i - \frac{1}{2}\beta t_i^2 \qquad (i = 1,2)$$

■

Proof: From Hypothesis 8.1 and Hypothesis 8.11, the price of products is $p_i = a - qi - \theta qi$. Meanwhile, from Hypotheses 8.3 and 8.11, the cost of products is known to be $c_i = C - kt_i - \delta kt_j$. So the profit of Enterprise i is given by

$$\begin{aligned}\pi_i &= pq_i - c_i q_i - \frac{1}{2}\beta t_i^2 \\ &= (a - q_i - \theta q_j)q_i - (C - kt_i - \delta kt_j)q_i - \frac{1}{2}\beta t_i^2 \\ &= (a - q_i - \theta q_j - C + kt_i + \delta kt_j)q_i - \frac{1}{2}\beta t_i^2\end{aligned}$$

❑

From Proposition 8.18, the profit of Enterprise 1 and that of Enterprise 2 could be obtained as follows.

$$\pi_1 = (a - q_1 - \theta q_2 - C + kt_1 + \delta kt_2)q_1 - \frac{1}{2}\beta t_1^2$$

$$\pi_2 = (a - q_2 - \theta q_1 - C + kt_2 + \delta kt_1)q_2 - \frac{1}{2}\beta t_2 \tag{8.62}$$

$$\frac{\partial \pi_1}{\partial q_1} = a - 2q_1 - \theta q_2 - C + kt_1 + \delta kt_2 = 0$$

$$\frac{\partial \pi_2}{\partial q_2} = a - 2q_2 - \theta q_1 - C + kt_2 + \delta kt_1 = 0 \tag{8.63}$$

Solving the equation,

$$q_1 = \frac{(\theta - 2)(a - C) + (\theta - 2\delta)kt_2 + (\delta\gamma - 2)kt_1}{\theta^2 - 4}$$

$$q_2 = \frac{(\theta - 2)(a - C) + (\theta - 2\delta)kt_1 + (\delta\theta - 2)kt_2}{\theta^2 - 4} \tag{8.64}$$

Substituting Formula 8.64 in Formula 8.62:

$$\pi_1 = \left(\frac{(\theta - 2)(a - C) + (\delta\theta - 2)kt_1 + (\theta - 2\delta)kt_2}{\theta^2 - 4}\right)^2 - \frac{1}{2}\beta t_1^2$$

$$\pi_2 = \left(\frac{(\theta - 2)(a - C) + (\theta - 2\delta)kt_1 + (\delta\theta - 2)kt_2}{\theta^2 - 4}\right)^2 - \frac{1}{2}\beta t_2^2 \tag{8.65}$$

when other conditions remain unchanged, the equilibrium, which is proportional to the spillover coefficient, is negatively correlated with the alternative coefficient.

Let $\frac{\partial \pi_1}{\partial t_1} = 2\frac{(\theta - 2)(a - C) + (\delta\theta - 2)kt_1 + (\theta - 2\delta)kt_2}{\theta^2 - 4} + \frac{(\delta\theta - 2)k}{\theta^2 - 4} - \beta t_1 = 0$

Then:

$$t_1^* = \frac{2(\theta - 2)(a - C) + 2(\theta - 2\delta)kt_2 + (\delta\theta - 2)k}{(\theta^2 - 4)\beta - 2(\delta\theta - 2)k} \tag{8.66}$$

And $\partial \frac{\partial^2 \pi_1}{\partial t_1^2} = \frac{(\delta\theta - 2)k + (4 - \theta^2)\beta}{\theta^2 - 4}$

$$\frac{\partial \pi_2}{\partial t_2} = 2\frac{(\theta - 2)(a - C) + (\theta - 2\delta)kt_1 + (\delta\theta - 2)kt_2}{\theta^2 - 4} + \frac{(\delta\theta - 2)k}{\theta^2 - 4} - \beta t_2 = 0$$

So:

$$t_2^* = \frac{2(\theta - 2)(a - C) + 2(\theta - 2\delta)kt_1 + (\delta\theta - 2)k}{(\theta^2 - 4)\beta - 2(\delta\theta - 2)k} \tag{8.67}$$

8.4.2 Numerical Analysis

Because the status of Enterprise 1 is equal to that of Enterprise 2, only numerical relations of Enterprise 1 are discussed. Unless otherwise mentioned, all the numerical values belong to Enterprise 1.

1. The relation between the equilibrium profit and the spillover coefficient of Enterprise 1

Assuming that $\alpha = 200$, $C = 10$, $k = 1$, $\beta = 100$, $\theta = 0.5$, $t_1 = 0.5$, $t_2 = 0.5$, δ rises from 0, in steps of 0.02, from Proposition 8.18, the relation between the equilibrium profit and the spillover coefficient, can be obtained as shown in Figure 8.20.

The bigger the spillover coefficient is, the more the equilibrium profit of Enterprise 1 will be, which can be concluded from Figure 8.20. Because Enterprise 1 not only gets profit from its innovation, but also from the innovation of Enterprise 2, the spillover of technology is beneficial to the receiver enterprise, which accepts the technology.

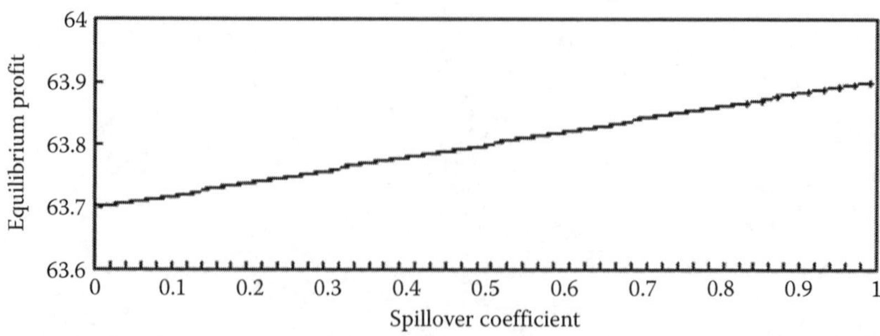

Figure 8.20 Relation between the equilibrium profit and the spillover coefficient.

2. The relation between the equilibrium profit and the alternative coefficient of Enterprise 1

Assuming that $\alpha = 200$, $C = 10$, $k = 1$, $\beta = 100$, $\delta = 0.5$, $t_1 = 0.5$, $t_2 = 0.5$, θ rises from 0, in steps of 0.02, from Proposition 8.18, the relation between the equilibrium profit and the alternative coefficient can be obtained, as shown in Figure 8.21.

The bigger the alternative coefficient is, the less the equilibrium profit of Enterprise 1 will be, which can be concluded from Figure 8.21. Because the products of Enterprise 1 could be substituted by the products innovated by Enterprise 2, the profit of Enterprise 1 may decrease because of the loss of some clients.

3. The relation between the optimal technical content and the spillover coefficient of Enterprise 1

It is assumed that $\alpha = 200$, $C = 10$, $k = 1$, $\beta = 100$, $\delta = 0.5$, $t_2 = 5$, θ rises from 0, in steps of 0.02. From Formula 8.66, the relation between the optimal technical content and the spillover coefficient can be deduced, as shown in Figure 8.22.

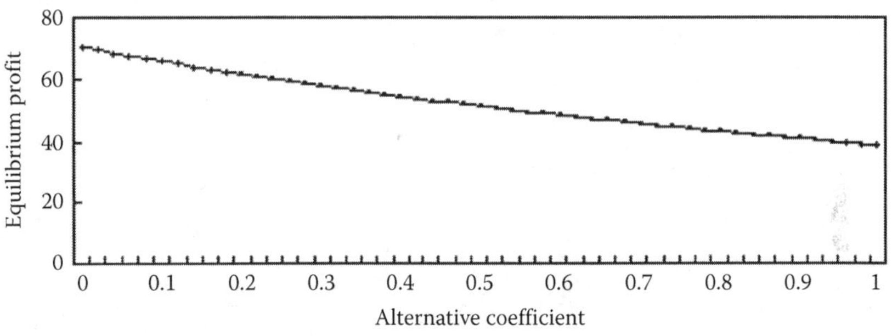

Figure 8.21 Relation between the equilibrium profit and the alternative coefficient.

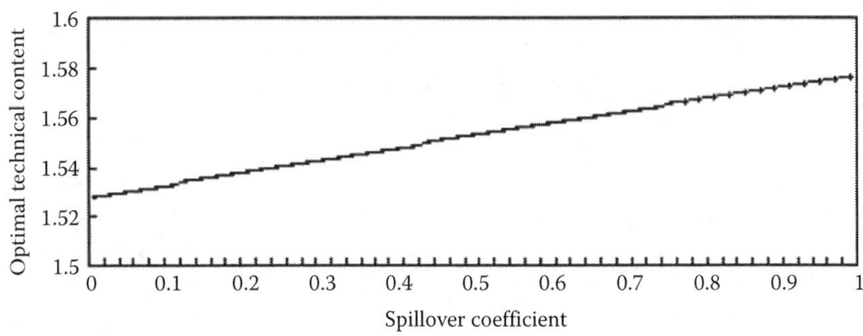

Figure 8.22 Relation between the optimal technical content and the spillover coefficient.

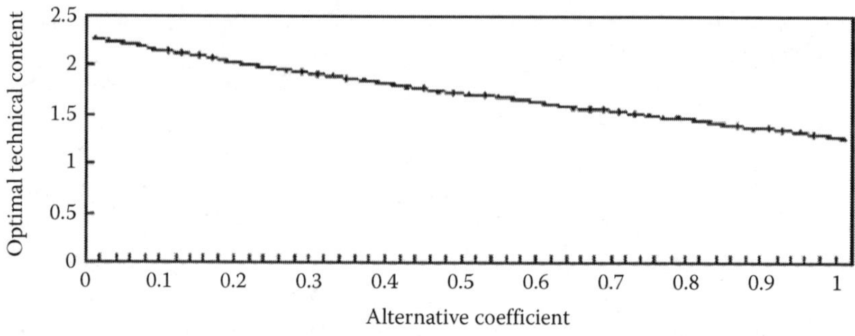

Figure 8.23 Relation between the optimal technical content and the alternative coefficient.

From Figure 8.22, it is clear that the optimal technical content is proportional to spillover coefficient. In other words, the bigger the spillover coefficient is, the more the optimal technical content of Enterprise 1 will be. Because the innovation of Enterprise 2 increases the technical content of Enterprise 1 besides its own, the spillover of technology is beneficial to enterprises, which receive technologies.

4. The relation between the optimal technical content and the alternative coefficient of Enterprise 1

It is assumed that $\alpha = 200$, $C = 10$, $k = 1$, $\beta = 100$, $\delta = 0.5$, $t_2 = 5$, δ rises from 0, in steps of 0.02. From Formula 8.66 the relation between the optimal technical content and the alternative coefficient can be deduced, as shown in Figure 8.23.

From Figure 8.23, it is clear that the optimal technical content is negatively correlated with the alternative coefficient. In other words, the bigger the alternative coefficient is, the less the optimal technical content of Enterprise 1 will be.

8.5 Evolutionary Game Model of Technical Innovation and Its Strategy Analysis

Game theory, taking the interaction of many decision subjects into consideration, mainly researches on a rational person taking decisions and the equilibrium problems of the decisions. The assumptions of the above analysis treat all enterprises as rational and the information as complete, which is too much for game subjects, as game theory cannot tell which equilibrium will be reached when facing too many choices. In real life, the decisions of enterprises, affected by the complicated changed environment, incomplete information of enterprises and some other aspects, may be irrational in the process of technical innovation. Evolutionary Games Theory succeeds in using groups to replace individuals in Game Theory in the following ways.

First, with the assumption of limited rationality, Evolutionary Games Theory gives the option of choosing decisions for Game subjects. Second, it treats groups as players instead of individuals. Finally, it substitutes ratio of different pure strategies selected by individuals in a group for a mixed strategy of players. This section will analyze the technical innovation of enterprise from the aspect of Evolutionary Games Theory.

8.5.1 Construction of the Model

In the evolutionary game problem of technical innovation, the enterprise has two strategies for selection, which are innovating and imitating. In the model, the decision is mainly a group decision and the behavior is organizational behavior. But participators' abilities to find the mistakes and adjust the strategy in time are poor. Furthermore, the changing economical behavior is a process of slow evolution, and not quick learning. As a result, in the process of dynamic replication of technical innovation, the rational level of participators is low. Each enterprise of the group should react (may not be optimal) to the game condition; they will either make technical innovation on their own or imitate other successful enterprises.

Taking the evolutionary game of technical innovation, with profit being uncertain as an example, its profit and loss value matrix is shown in Table 8.1.

In evolutionary game problems, the rational level and learning speed of the members is low. We make random pairs of members of the group, and the game repeats. Low learning speed means the transformation towards optimal strategy is not realized by all game players at the same time, but gradually realized in different times. Assuming that, in Game 1, ratio x of individuals takes the strategy of innovation, and ratio $1 - x$ of them takes the strategy of imitation. The expected profit of Game i is u_{ij} $(i, j = 1, 2)$ when choosing the strategy of j, and the average profit of Game i is $\overline{u}_i$. Then the expected and the average expected profit of Game 1 could be calculated as follows.

$$\begin{aligned} u_{11} &= ya + (1 - y)c \\ u_{12} &= ye + (1 - y)g \end{aligned} \tag{8.68}$$

$$\overline{u}_1 = xu_{11} + (1 - x)u_{12} \tag{8.69}$$

Table 8.1 Profit and Loss Value Matrix of 2 × 2 Asymmetric Game

Player-2		*Innovation*	*Imitation*
Player-1	Innovation	*a*, *b*	*c*, *d*
	Imitation	*e*, *f*	*g*, *h*

The expected profit and average expected profit of Game 2 are as follows:

$$u_{21} = xb + (1-x)f$$

$$u_{22} = xd + (1-x)h \tag{8.70}$$

$$\bar{u}_2 = yu_{21} + (1-y)u_{22} \tag{8.71}$$

According to the ideology of evolutionary replicator dynamics of biology, the game player who benefits less than the others will change his strategy so as to get more profit. Then the ratio of different strategies being selected will change. Meanwhile, the changing rate of special strategy's ratio is proportional to the range between its profit and average profit. So the changing speed of ratio x of player 1 taking the strategy of innovation could be represented as the following differential equation.

$$\begin{aligned}\frac{dx}{dt} &= x(u_{11} - \bar{u}) = x(u_{11} - xu_{11} - (1-x)u_{12}) \\ &= x(1-x)[(a-e+g-c)y + c - g]\end{aligned} \tag{8.72}$$

Let Formula 8.72 be equal to 0, and the equilibrium point of the replicator dynamics equation can be obtained. In this situation, ratio x of player 1 taking the strategy of innovation is stable in the process of replicator dynamics.

$$x_1 = 0, \quad x_2 = 1, \quad y = \frac{g-c}{a-e+g-c}$$

The changing speed of ratio y of player 2 taking the strategy of innovation could be represented by the following differential equation.

$$\begin{aligned}\frac{dy}{dt} &= y(u_{21} - \bar{u}_2) = y(u_{21} - yu_{21} - (1-y)u_{22}) \\ &= y(1-y)[(b-d+h-f)y + f - h]\end{aligned} \tag{8.73}$$

Let Formula 8.73 be equal to 0. In this situation, ratio y of player 2 taking the strategy of innovation is stable in the process of replicator dynamics.

$$y_1 = 0, \quad y_2 = 1, \quad x = \frac{h-f}{b-d+h-f}$$

The five equilibrium points calculated are

$$A(0,0),\quad B(0,1),\quad C(1,0),\quad D(1,1),\quad E\left(x=\frac{h-f}{b-d+h-f},\ y=\frac{g-c}{a-e+g-c}\right)$$

The first four stable points imply members of the group tend to take the same strategy (1 or 2). Point A implies player 1 and player 2 both take the strategy of imitation; Point B implies player 1 takes the strategy of imitation whereas player 2 takes the strategy of innovation; Point C implies player 1 takes the strategy of innovation whereas player 2 takes the strategy of imitation; Point D implies player 1 and player 2 both take the strategy of innovation; Point E implies different members of the group take different strategies in some definite ratio. The first four points are corresponding to the pure strategy equilibrium of the perfect rational game, and the last point corresponds to the mixed strategy equilibrium. It is noteworthy that these stable points only mean that the ratio of players taking special strategy have reached a certain level, which will not change, but which stable point the process of replicator dynamics will tend to cannot be predicted. Depending on the specific problems, the tendency will be decided by the initial status of the ratio of players taking strategies and the condition of the dynamic differential equation in the corresponding interval.

Besides, real stability should be of anti-interference. That is to say, if the above proportion relation deviates from these stable points, replicator dynamics may reset it. In addition the derivative of stable points should less than 0; in other words, the slope of the tangent line should be negative. If the point (x, y) satisfies the above requirements, it can be called the evolutionary stable strategy in the evolutionary game, which means players with limited rationality have stable proportions in the process of dynamic strategy adjustment.

8.5.2 Case Study

It is assumed that player 1 and player 2 are competitive enterprises in the game. Their initial profits are respectively 160 and 150. If one of them carries out innovation, the market share of the other one will decrease. When player 1 carries out technical innovation, their profits will change to 200 and 60, respectively; when player 2 carries out technical innovation, their profits will change to 70 and 210; when both the enterprises make innovative investigation, their profit will decrease to 60 and 40 respectively after deducting the innovation cost. The loss and profit matrix of this Game is shown in Figure 8.2 and Table 8.2.

It is clear from Figure 8.2 that the advantages of innovation are many. When player 1 carries out innovative investigation, choosing the strategy of imitation is optimal to player 2. However, will player 2 wait for its doom? No. Every enterprise

Table 8.2 Loss and Profit Matrix

Player-2		*Innovation*	*Imitation*
Player-1	Innovation	60,40	200,60
	Imitation	70,210	160,150

will take existing competition into account, and the greatness of player 1 is tantamount to the doom of player 2, so it is likely that player 2 will make innovative investment at the same time.

Above all, the evolutionary stable strategy of this model is $E(0.75,0.8)$. In this strategy, three-fourths of the enterprises at player 1 level innovate, and the others carry out imitation. Meanwhile, four-fifths of the enterprises at player 2 level innovate and the others carry out imitation.

8.5.3 Results Analysis

From the evolutionary games model, we can see that, in the long-term innovative evolution of the big-scale group, when the profit of enterprises after innovation is according to the above assumption, the number of individuals at the player 1 level taking the strategy of innovation will be stable at 75 percent, while the others carry out imitation. The number at the player 2 level will be 80 percent. Because innovation requires more investment, in a certain period the profit gained by innovation cannot catch up with that gained by imitation. However, even though the current benefit is not satisfactory, developing the market and founding a brand by means of innovation may be more attractive for most enterprises. In a word, the economic benefit of innovation is more than that of imitation in a long-term goal, in terms of feasibility.

Chapter 9

Comprehensive Measurement of Technology Transfer and the Analysis of Its Impact on Economic Growth

This chapter focuses on the hierarchical structure and issues of the comprehensive measure of technology transfer, and how they affect the causal relationship between China's economic growth and technology transfer through structural indicators of the degree of technology transfer and the Granger causality model of technology transfer. This chapter analyzes the spillover effect of Foreign direct investment (FDI), which is an important aspect of international technology transfer domestically and the spillover effects of FDI, and China's technology in promoting technological progress and economic growth. FDI, accompanied by advanced production technology and advanced management methods and technology spillover effect, directly promotes China's technological progress and economic growth, improves the competitiveness of China's enterprises and products in the international market, and this will be beneficial to China's economic growth in both the short- and long-term.

9.1 Comprehensive Measure of Technology Transfer

9.1.1 International Technology Transfer Measurement

The indicators of the measure of international technology transfer are technology balance of payments (the difference), the license revenue and expenditure, patent rate of self-reflection, the rate of technology development locally, FDI, scientific publications and reference, scientific movement and the number of scientific colloquiums, etc., as shown in Table 9.1. And the three technical indicators—technology balance of payments (the difference), the license payments, and patents rate—are direct measures for the number of international technology transfers, and the others are the indirect indicators added as supplementary ones.

Some of the indicators are explained here in brief:

1. International technology balance

Technical indicators for the balance of payments were first developed by the organization for economic cooperation and development (OECD). They serve as a measure of scientific and technology activity in the second generation system on the manual and basic skills in OECD member countries to promote the use of statistics.

OECD's technology balance of payments is defined to measure the flow of international technology through records of interstate transactions relating to the intangible trade and the invisible trade, which has something to do with the technical knowledge and the trade-related services including technical content. Technology balance of payments is just an integral part of international technology transfer;

Table 9.1 The Indicators of the Measure of International Technology Transfer

The three main indexes	1. International technology balance of payments (the difference)
	2. Permit payments
	3. The rate of technological self-satisfaction
Supplementary indicators	1. Foreign direct investment (FDI)
	2. The patent self-reflection rate
	3. Technical publications and citation
	4. Mobile technology workers and the number of meetings to discuss the technical issues

therefore, other data should also be added to international technology transfer, such as FDI and so on.

2. The rate of technical self-satisfaction

International technology dependence is an important indicator of domestic dependence on foreign technology, which can be indicated as the trade balance, which is the ratio of the export of technology to the introduction of technology, or can be measured as the ratio of the introduction of technology to the expenditure for R&D. In general, the higher the balance of technology trade, the higher the technological competitiveness.

The rate of technical self-satisfaction reflects the domestic technological standard for meeting the demand, according to the formula:

Technical self-satisfaction rate = 1 – the degree of dependence on international technology

Technical self-satisfaction rate = 1 – the technology trade balance

3. Patent rate of self-reflection

Patents can reflect the innovative activities of enterprises, intellectual property ownership, and the level of technological development, technical and economic competitiveness in both the macro and micro levels. Generally speaking, a country's international patent cases can be an important indicator of its technical competitiveness characterization. Methods of protecting intellectual property rights have become increasingly more effective worldwide. In the international competition, any enterprise in one country must declare its patent in other countries to maintain its place in international competition and if it has an ambition to explore the international market with its own unique technology. The quantity and quality of patents of an enterprise in foreign countries, to a certain extent, indicate its international competitiveness.

The patent rate of self-reflection reflects a country's demand for foreign patents. The formula for the measure is shown as follows:

Patented self-reflection rate = number of patents registered by citizens/number of patents registered domestically.

The total number of patents domestically registered = number of patents registered by citizens and the number of patents registered by foreigners to apply for authorization

4. License revenue and expenditure

There are important differences between license revenue and expenditure and international technology balance which refers to the invisible authority right of revenue and expenditure (such as patents, copyrights, trademarks, and franchising). License

Table 9.2 Chinese Technology Transfer Indicators Measure

Technology Transfer Level	*Indicators*
International technology transfer	1. Foreign direct investment
	2. The introduction of foreign technology
	3. High-tech product imports
	4. Foreign economic cooperation
	5. High-tech product exports
Domestic technology transfer	1. Domestic market turnover of technology
	2. Appreciation in domestic knowledge-based services

revenue and expenditure are covered by a clear, direct range, which reflects the level of technology transfer in these forms of ownership.

9.1.2 Technology Transfer Indicators of Measurement in China

As a system, technology transfer has a hierarchical structure; therefore, in establishing the measure of technology transfer level indicators in China, the subindexes corresponding to every subsystem are taken into account, and then integrated to reflect the level of China's comprehensive technology transfer.

The caliber of the statistical data is inconsistent for China's measurement of technology transfer since China's system of indicators does not cover the OECD indicator system. To measure the level of China's comprehensive technology transfer, indicators are called for to adapt to China's own situation. According to the available resources in China, taking into account the availability of data in such areas, China's technology transfer can be divided into international transfer and domestic transfer, as shown in Table 9.2.

9.1.3 Indicators Analysis of Measurement in Technology Transfer

9.1.3.1 FDI Technology Spillover

After the reform and the opening up, China has made great efforts to attract foreign investment to such a large extent FDI is flooding into China. FDI has had a very great impact on technology progress in China. So FDI is considered an important indicator of international technology inflow, introducing FDI as an indicator in the comprehensive measurement of technology transfer.

9.1.3.2 Introduction of Foreign Technology (JSYJ)

1. An overview of the introduction of foreign technology

JSYJ is used as the symbol for the introduction of foreign technology. The introduction of technology denotes the transfer of technology from abroad to home. Specifically, the technology is a means of bringing foreign technology knowledge and experience, as well as the necessary ancillary equipment and apparatus for the development of the national economy to make scientific and technological progress. Technology introduction contains the following three concepts:

First, the technology introduction is introduction of foreign technologies to the domestic market.

Second, there are principal differences between technology introduction and import of equipments. People often divide technology into software and hardware technologies in a broad sense. Software technology is the aforementioned technical knowledge, experience, and skills, being a kind of 'pure' technology. Hardware technology, on the other hand, refers to materialized equipment like physical and chemical techniques. The import of equipment is commonly understood as people merely purchasing equipments from abroad instead of software technology. If people purchase software technology from abroad with some equipment, it can then be regarded as the introduction of technology.

Third, the introduction of technology is designed to improve the manufacturing capacity, the level of technology and management of one nation or an enterprise, only by introducing the technology of the software and absorbing it.

2. The description of China's recent introduction of foreign technology

China's industry phylogeny is the history of the introduction of foreign technology in some sense. After the reform and the opening up, domestic industry was staggered by the advanced science and technology of the developed countries. Therefore, the introduction of foreign technology was regarded as a shortcut to enter the advanced world of technology and to alter the disadvantageous status of China's industrial technology. Especially since the 1990s, Chinese economy has grown at great speed continually and come up with significant achievements, which have something to do with the positive Chinese attitude to the introduction of foreign technology. From Table 9.3, we can see that technology sales have increased year by year from 1994 to 2000. In 1994 the introduction of foreign technology was worth \$ 4,105,760,000, whereas, in 2000 it amounted to \$ 18,175,960,000, an increase of 342.69 percent. Soon after, the country recognized that blind introduction of technology resulted in attaching too much importance to introduction rather than the digestion of the technology, emphasizing hardware instead of software, introducing capital instead of technology, and a very serious phenomenon of duplicate introduction, resulting in huge numbers of technology import contracts since 2000.

Table 9.3 Level of Development of Introduction of Foreign Technology (Unit: 10,000 $) along with Growth Year by Year

Year	*JSYJ*	*Growth Rate*	*Year*	*JSYJ*	*Growth Rate*	*Year*	*JSYJ*	*Growth Rate*
1991	345,923	100 percent	1996	1,525,700	17.07 percent	2001	909,090	–49.98 percent
1992	658,988	90.50 percent	1997	1,592,312	4.37 percent	2002	1,738,920	91.28 percent
1993	610,943	–7.29 percent	1998	1,637,510	2.84 percent	2003	1,345,121	–22.65 percent
1994	410,576	–32.80 percent	1999	1,716,221	4.81 percent	2004	1,385,558	3.01 percent
1995	1,303,264	217.42 percent	2000	1,817,596	5.91 percent			

Source: National Bureau of Statistics of China, *Chinese Statistical Yearbook*, (*2005*). China Statistical Press, Beijing, IO, 2006.

It is obvious that China's technology import contracts have remained stagnant. There are many reasons, not only foreign but also domestic, for this. The main internal factor is the shift of Chinese technology policy from mere introduction to the combination of digestion and absorption along with innovation.

9.1.3.3 High-Tech Product Imports (GJSJK)

1. The definition of high-tech products

GJSJK is used as the symbol of high-tech product imports. It is recognized by OECD, the United States and other countries that "high technology" refers to new industrial technology or cutting-edge technologies, which have a huge impact on the military functions and the economy of the whole country and have even greater social significance.

The concept of "high-tech," varies with time and location. As a result, high-tech products should be defined according to the different stage characteristics of economic development as far as possible to enable international comparisons using international methods.

At present, the standard that measures high-tech products is not perfect, and as yet the unified standard measure of high-tech products has not been established internationally. As the high-tech products are closely related to the level of economic development, at a new stage of economic development, the original high-tech products may change into "old" products, with the emergence of possible new products. In addition, due to the uneven national economic development, in a certain period, the mature industrial products in developed countries may be high-tech products in developing countries. Therefore, it is a very complex task to define high-tech products. Nevertheless, the high-tech products are still defined as follows with three main indicators: first, R&D intensity; second, the proportion of scientific and technology talents; and third, the technical complexity of the products, such as the technological level of products, the equipment for producing products, and the level of technology, etc.

2. Recent situation in China's high-tech product imports

In 2005, the expenditure on China's imported high-tech products amounted to \$197,710,000,000, accounting for 30.0 percent of the total share, which is higher than the level in the same period of the previous year. In 2005 the expenditure on China's imported high-tech products increased by 22.57 percent over the same period last year, but the growth rate dropped by 13 percentage points in 2004 as shown in Table 9.4. In 2005, China's high-tech imported products came mainly from ASEAN, accounting for 19.7 percent of China's total imported high-tech products, followed by Japan, Taiwan of China and South Korea, the accounts being respectively 15.4, 14.8, and 13.1 percent, while European Union and United States were respectively 9.5 percent and 8.1 percent. This shows that the United States,

Table 9.4 Development Level of High-Tech Product Import (in: $ 1,000,000) along with Growth by Years

Year	*GJSJK*	*Growth Rate*	*Year*	*GJSJK*	*Growth Rate*	*Year*	*GJSJK*	*Growth Rate*
1991	9,439	100 percent	1996	22,469	2.94 percent	2001	64,107	22.09 percent
1992	10,712	13.49 percent	1997	23,893	6.34 percent	2002	82,800	29.16 percent
1993	15,909	48.52 percent	1998	29,201	22.22 percent	2003	119,300	44.08 percent
1994	20,595	29.46 percent	1999	37,598	28.76 percent	2004	119,300	35.21 percent
1995	21,827	5.98 percent	2000	52,507	39.65 percent	2005	197,710	22.57 percent

Source: National Bureau of Statistics of China, *Chinese Statistical Yearbook, (2005)*. China Statistical Press, Beijing, IO, 2006.

European Union, Japan, and other developed countries remain the most important consumer markets of Chinese high-tech products, while China's import trade partners are developed widely. Meanwhile, the newly industrialized countries as well as the adjacent developing countries began to seize markets of China's high-tech products with their more powerful and competitive advantages.

9.1.3.4 Foreign Economic Cooperation (DWJJHZ)

1. The content of external economic cooperation

In a narrow sense, the foreign economic cooperation (DWJJHZ) denotes contracted projects and labor services of Chinese enterprises in foreign countries after being approved by China's relevant departments. In a broad sense, Chinese foreign economic cooperation includes external economic and technical aid; borrowing of foreign funds, foreign investment; foreign contracted projects, and labor cooperation; foreign joint ventures or enterprises with Chinese ownership; foreign production technology contracts; multilateral cooperation with the United Nations Development System and other international organizations; economic and technological aid from friendly countries.

2. The outline of Chinese economic cooperation and development

As can be seen in Table 9.5, from 1991 to 2004, the contracts of foreign economic cooperation have been increasing year after year. In 1991, the sale of foreign

Table 9.5 Level of Development of Economic Cooperation with Foreign Countries (Unit: Billion $) along with Growth by Year

Year	*DWJJHZ*	*Growth Rate*	*Year*	*DWJJHZ*	*Growth Rate*	*Year*	*DWJJHZ*	*Growth Rate*
1991	36.09		1996	102.73	6.21 percent	2001	164.55	10.12 percent
1992	65.85	82.46 percent	1997	113.56	10.54 percent	2002	178.91	8.73 percent
1993	68	3.26 percent	1998	117.73	3.67 percent	2003	209.3	16.99 percent
1994	79.88	17.47 percent	1999	130.02	10.44 percent	2004	276.98	32.34 percent
1995	96.72	21.08 percent	2000	149.43	14.93 percent			

Source: National Bureau of Statistics of China, *Chinese Statistical Yearbook*, (*2005*). China Statistical Press, Beijing, IO, 2006.

economic cooperation contracts amounted to $3,609,000,000, whereas in 2004, it amounted to $ 27,698,000,000. So the cumulative growth is 667.47 percent. In economic cooperation with foreign countries, foreign project contracting is the most important part, which accounts for more than 80 percent of the total contracts. In view of such a trend, a growing number of consulting services for design has been integrated into the foreign engineering contract projects.

9.1.3.5 High-Tech Exports (GJSCK)

GJSCK is used as the symbol of high-tech exports. In recent years, China has seized significant opportunities in the new trend in global reconstruction and optimization of productive elements as well as transfers of productivity. At the same time, it actively implements the strategy of rejuvenation in science and education and promotes the rapid development of high-tech industry. High-tech industry has played an increasingly prominent role in promoting industrial restructure and development of foreign trade. China's high-tech exports show the following characteristics.

First, the growth of high-tech exports has been increasing very fast. Since the 1990s, China's high-tech exports have maintained a rapid growth momentum and the proportion of the exports has been rising. In 1991, the sale of Chinese high-tech product exports amounted to $ 2,877,000,000, whereas in 2005, the sale amounted to $ 218,250,000,000, an increase of 74.86 times, and the average annual growth rate was 36.93 percent, as seen in Table 9.6.

Second, high-tech product export competitiveness has gone up gradually with gradual increases in the TSC index. China's high-tech product export

Table 9.6 Level of Development of High-Tech Exports (Unit: Million $) along with Growth by Year

Year	*GJSCK*	*Growth Rate*	*Year*	*GJSCK*	*Growth Rate*	*Year*	*GJSCK*	*Growth Rate*
1991	2,877		1996	12,663	25.49 percent	2001	46,452	25.40 percent
1992	3,996	38.89 percent	1997	16,310	28.80 percent	2002	67,900	46.17 percent
1993	4,676	17.02 percent	1998	20,251	24.16 percent	2003	110,300	62.44 percent
1994	6,342	35.63 percent	1999	24,704	21.99 percent	2004	165,400	49.95 percent
1995	10,091	59.11 percent	2000	37,043	49.95 percent	2005	218,250	31.95 percent

Source: National Bureau of Statistics of China, *Chinese Statistical Yearbook*, (2005). China Statistical Press, Beijing, IO, 2006.

competitiveness can be measured by the TSC index, which reflects the degree of intra-industry trade competitive advantage.

TSC index for the basic formula is shown as follows:

$$\text{TSC}_i = (E_i - I_i)/(E_i + I_i) \in \tag{9.1}$$

where

TSC_i is the trade specialization index for the product i

E_i and I_i represent export and imports for product i, respectively

The closer the TSC is to 1, the stronger the international competitiveness is, while the closer the TSC is to –1, the weaker the international competitiveness is.

9.1.3.6 Technology Market Turnover

Technology market is an integral part of the socialistic market system. The development of technology market not only changes and enriches the commodity market economy and expands the production material market, but also speeds up the expansion of the financial markets, labor market, information market, and markets of other factors. Technical achievement transfer functions primarily in the technology market, which builds a bridge to link research institutions and enterprises, bringing them close to each other. The technology market is gradually changing from the exploitative market to the managerial market, which has recently become the main market for Science–Industry Trade and Science–Agriculture Trade. At the same time, the technology market promotes technological progress, while the enterprises absorb, assimilate a new technique in the domestic technology market, and carry out technological transformation, product development, and technology portfolio development, resulting in increase in its technical capacity. Enterprises have become the largest buyers in the technology market, and have gradually been playing a very active role in the market.

From 1992 to 2004, the turnover in China's technology market increased year by year, and T is used to stand for the technology turnover. In 1992, the turnover of the Chinese technology market amounted to ¥ 14,218,300,000, while in 2004 it was ¥ 123,737,190,000; during a 13-year period the cumulative growth amounted to 770.27 percent, as shown in Table 9.7.

Over the past decade, China's technology market has made a great effort in developing legislation, organization hierarchy, management, operating systems, and operating mechanisms, etc. Because the technology market has become an important approach to transfer of technology achievements, the relationship between the technology market and the market of related factors is enhanced gradually. Technology market services are developing in a diversified and comprehensive manner.

Table 9.7 Level of Development in Technology Market Turnover (Unit: Million $) along with Growth by Year

Year	*T*	*Growth Rate*	*Year*	*T*	*Growth Rate*	*Year*	*T*	*Growth Rate*
1992	1,431,830	–	1997	3,513,718	17.04 percent	2002	8,106,455	6.27 percent
1993	2,075,540	44.96 percent	1998	4,358,228	24.03 percent	2003	10,136,031	25.04 percent
1994	2,288,696	10.27 percent	1999	5,177,304	18.79 percent	2004	12,373,719	22.08 percent
1995	2,683,447	17.25 percent	2000	6,250,334	20.73 percent			
1996	3,002,045	11.87 percent	2001	7,627,983	22.04 percent			

Source: National Bureau of Statistics of China, *Chinese Statistical Yearbook*, (*2005*). China Statistical Press, Beijing, IO, 2006.

9.1.3.7 Soft Technology Spillover in the Knowledge-Based Service Sector (Z)

1. The academic understanding of the concept of soft technology

It is generally agreed in the academic field that soft technology is developed with soft science, but there are various interpretations of soft technology. Currently, Chinese scholars have certain views which can be generalized in four perspectives as follows:

a. In the application of soft science

Hongxia refers to the application of soft science methods as "soft technology methods," such as the Delphi approach, brainstorming, Gordon Law, etc. The Xixing Men and other scholars use the relationship between soft technology and soft science to explain the meaning of soft technology. Tsang Tak-tsung considers the so-called soft technology as a scientific system about decision-making and a multidisciplinary group consisting of various subjects, such as management science, system science, scientific decision-making forecasts, and so on, which is also the product of general science, large-scale production, and general economic development.

b. Understanding the dualism of soft technology

In dualism on technology, the technology will be divided into two elements; one is the hard element, whereas the other is the soft element. According to the technical material form, many scholars call the technology with the material form as hard technology, and the others are called soft technology. In the sense of technology, Yulin Liu and Jie Zhao put forward the concept of soft and hard technologies. Soft technology includes technology transfer, technology licensing, technology services, and technical advisory. Ning Shu regards the improvement of knowledge, the impact of policies and laws, etc. as soft technology progress, which is divided according to the hardness of technology and the progress of the materialized technology as hard technology advances, such as variation in the quality of production elements, etc. This view is based on the properties of the technical carrier to analyze soft technology. Technology with the material carrier is called hard technology and the other is called soft technology.

c. Understanding the meaning of soft technology according to managerial techniques

In understanding the soft technology from the aspects of managerial techniques, there are controversial points of view. One holds that soft technology is equivalent to management and its techniques. For instance, Yongling Zhong, equates the promotion of soft technology progress to raising the level of management in the perspective of technical progress. Xiangqian Zhang, in a discussion on the relationship between management and knowledge-based economy, deems soft technology as general managerial skills. Another view holds that soft technology includes soft managerial techniques. For example, Guohong Chen deems that the breakthrough in soft technology is primarily reflected in the variation of organizations, management, and decision-making skills in the analysis of the relationship between technological innovation, technology transfer, and progress. There are other scholars, such as Yiquan Xing and so on, who claim that soft technology is merely managerial skills or a combination of organization, decision-making, and management.

d. Understanding soft technology from a broader perspective

Domestic scholars, like Zhengying Zhou, interpret soft technology in a broader sense. Zhengying Zhou uses Plato's understanding of technology as the origin of the theory of analysis of soft technology. He reckons that soft technology is an intelligence technology, innovated in the domain of people, human behavior, and human society. Soft technology is founded or summarized as a principle in economic, social, and cultural activities and comprises a system of rules, regulations, mechanisms, methods, and procedures to solve problems on the basis of improvement, adaptation, and control in the objective and subjective world.

Apart from these four perspectives in understanding the concept of soft technology, there are some other scholars who put forward other views. For instance, Yanlai Chen equates technology with experience, ability, and consciousness; Zhengrong She and Yunfei Zhang consider soft technology as equal to technology of ecological environmental protection. Consequently, there are great differences in the understanding of the concept of soft technology.

2. The connotation as a knowledge-based service industry

Z stands for the soft technology spillover in the knowledge-based service. With regard to knowledge-based service, different scholars have different definitions. Miles (1995) defined the knowledge-based services as enterprises and organizations, which depend on professional knowledge and provide knowledge-based intermediary products and services. Hauknes (1996) thinks that knowledge-based service industries are technology-intensive and information-oriented service industries. From the study, it is clear that the knowledge-based service industry comprises general enterprises that provide knowledge-based services, like legal services, management consulting services, accounting services, financial services, computer and information services, and engineering design services. Some scholars believe that the knowledge-based service industries are knowledge-intensive industries engaged in the proliferation of knowledge, achievements transfer, technology assessment, technology transfer, investment and financing services, information consulting, management consulting, and training, etc.

3. The role of knowledge-based service industry in China's economy

The knowledge-based service industry is the source of knowledge and the transfer node of the knowledge-based innovation system, which affects the performance in the innovation system primarily through the development of new technologies, promoting the flow of knowledge, changing the way of innovation, and enhancing the portfolio of innovations.

First, knowledge-based service industry promotes the development of new techniques. There is a positive feedback relationship between new technologies and services. New techniques expand the range of knowledge-based services, while new services promote the technology through research, design, and engineering. The emergence of biotechnology, new materials, environmental protection technology, and ICT technology expands the knowledge-based service sector and promotes its innovation, and in turn the innovation and service sector of knowledge-based services accelerate these new techniques and experimental production.

Second, the knowledge-based service industry promotes the flow of knowledge, that is, the flow of "soft" technology. The competition advantage is closely related to innovation network and the location of the resource of knowledge. The knowledge-based service industry increases the adaptability of the enterprises, promotes the flow of personnel, and provides the concepts of knowledge, information, and management as well as technological innovation of customers through complementary expertise. The knowledge-based service industry transfers innovative ideas to customers and improves the innovation to meet different demands, which promotes the flow of the knowledge among different customers. The impact of knowledge-based innovation on the services sector depends on the effective service, the closer and the more durable knowledge-based service industry connections with customers, the more likely the organizational and technical skills will combine with the users' innovation strategy and the greater the impact of innovation on customers.

Third, the knowledge-based service industry changes the way of innovation. Antonelli, who studied the interactive relationship between users and the knowledge-based service industry, found that the vertical structure of knowledge, established on the basis of laboratory research after the war, had been replaced by the interaction of the market in reality, online users, and producers of knowledge. The knowledge-based service industry transfers knowledge within innovation networks and among innovation networks, which have become important nodes of knowledge flow in the innovation network. ICT technology has changed the traditional flow of knowledge, promoting business organizations to change from the traditional pyramid structure to the flat and flexible organizational structure, which is an important basis for absorption of knowledge, expanding the time and space of the knowledge-based service industry.

Finally, the knowledge-based service industry strengthens the combination of innovation. Effective technological innovation must be accompanied by market innovation, staff development, and organizational innovation. Knowledge-intensive service industries will not only have an impact on innovation in technology but also on the interaction with organizations of customers, relationship, and market.

4. The selections of the typical knowledge-based service industry

In November 2003, the National Bureau of Statistics began to reform the caliber of statistical data collection and analysis. The main elements of the reform are to establish statistics in the knowledge-based service industry, which did not exist before. From the current *China Statistical Yearbook* in the classification of the tertiary industry, the following four sectors are chosen on behalf of China's knowledge-based service industry: telecommunications, real estate, scientific research and comprehensive services, and educational media.

5. Spillover of soft technology in the typical knowledge-based service sectors

Zhengying Zhou said not only is technology in the traditional sense classified as part of the soft technology, but also the operational part of social science and nonscientific knowledge, which was separated into soft areas of technology and operational knowledge, never before considered as techniques, also belong to soft technology. In his opinion, service is a process technology and belongs to the category of soft technology.

Based on the analysis of the connotation of a knowledge-based service industry from above, it is evident that knowledge-based service industry belongs to the category of soft technology, resulting in the spillover of soft technology from the industry, which provides knowledge-based services, accordingly leading to soft technology transfer.

The communication industry provides the customers with a full range of innovative products to meet their demands, services, and solutions and enables customers to obtain problem-solving abilities through the delivery in this kind of service, where the aim is to achieve the flow of soft technology from communication industry to customers.

As different from the architecture industry, which belongs to the second type, real estate is regarded as a service industry. Real estate is an integrated industry engaged in design, planning, consulting management, marketing management and service. These comprehensive services transfer among companies and from companies to individuals, making real estate services a means of the soft technology transfer.

Scientific research industry and comprehensive technical service industry are, in the national economy, two major categories according to the "code of the national economy and industry." The scientific research industry can be divided into natural sciences, social sciences, and other scientific research, and the integrated technical service industry can be divided into services for technical promotion and scientific communication, engineering design industry, and other integrated technical service sectors. Scientific research and comprehensive technical service can achieve the flow of soft technology through the publication of scientific research achievements and comprehensive technical services.

Educational media industry provides services and also transfers innovative ideas to customers, and alters these innovations to meet the demands of different customers, promoting the knowledge of technology in the flow of soft technology among different customers. The closer and the more durable the knowledge-based service industry's connections with customers, the higher the possibility that the organizational and technical skills will combine with the customers' innovation strategies, resulting in a greater impact.

6. The measurement of spillover in soft technology in the typical knowledge-based service sectors

The spillover effects of the soft technology are very obvious in a typical knowledge-based technology. The development of the knowledge-based service effectively accelerates soft technology transfer, helping the progress of China's technology.

Here we choose the four sectors of the knowledge-based service industry in China, and use the added value of these four sectors as the measurement of the level of soft technology spillover, which is brought into the indicator system of the Chinese technology transfer measurement.

Table 9.8 represents the value-added data in the knowledge-based service industry. As an important carrier of technology transfer, the continual development

Table 9.8 Added Value in Knowledge-Based Services (Unit: Billion ¥)

Year	*Z*	*Growth Rate*	*Year*	*Z*	*Growth Rate*	*Year*	*Z*	*Growth Rate*
1992	1387.6	—	1997	4373.7	17.98 percent	2002	8705.5	13.59 percent
1993	1802.1	29.87 percent	1998	4982.4	13.92 percent	2003	9889.7	13.60 percent
1994	2542.9	41.11 percent	1999	5585.2	12.10 percent			
1995	3136.9	23.36 percent	2000	6703	20.01 percent			
1996	3707.3	18.18 percent	2001	7664.2	14.34 percent			

Source: National Bureau of Statistics of China, *Chinese Statistical Yearbook, (2005)*. China Statistical Press, Beijing, IO, 2006.

of knowledge-based service sectors indicates that increasingly more attention is paid to soft technology. As soft technology spreads out with the steady development of the knowledge-based service industry, the knowledge-based service industry has become an important medium of soft technology transfer.

9.1.4 Chinese Technology Transfer Degrees

1. Definition of degrees of technology transfer in China

Based on the Chinese technology transfer system over the target, an indicator called the degree of technology transfer is defined, which reflects the general level of technology transfer in China. The definition is shown as follows:

$$\mathrm{MTF} = \frac{\mathrm{T} + \mathrm{Z} + \mathrm{FDI} + \mathrm{JSYJ} + \mathrm{GJSJK} + \mathrm{DWJJHZ} + \mathrm{GJSCK}}{\mathrm{GDP}} \tag{9.2}$$

MTF, T, Z, FDI, JSYJ, GJSJK, DWJJHZ, DWJJHZ, GJSCK, and gross domestic product (GDP) respectively stand for China's technology transfer degrees, the Chinese technology market transactions, the amount of value-added knowledge-based services, FDI, the introduction of foreign technology, high-tech imports, foreign economic cooperation, high-tech exports, and China's gross domestic product.

2. Chinese technology transfer calculations
   a. The process of data

Prior to calculating China's technology transfer degrees, these indicators of the Chinese technology transfer are converted with the average annual exchange rate of RMB against the U.S. dollar, seen in Table 9.9; and the data are amended to eliminate the effect of price. The price-amending indexes here are CPI, seen in Table 9.10.

Table 9.9 Average Annual Exchange Rate Under the Price of RMB against the U.S. Dollars in Indirect Method

Year	*1992*	*1993*	*1994*	*1995*	*1996*	*1997*	*1998*
X	551.46	576.20	861.87	835.10	831.42	828.98	827.91
Year	1999	2000	2001	2002	2003	2004	
X	827.83	827.84	827.70	827.70	827.70	827.68	

Source: National Bureau of Statistics of China, *Chinese Statistical Yearbook*, (*2005*). China Statistical Press, Beijing, IO, 2006.

Note: *X* stands for exchange rate.

Table 9.10 CPI Index (Chain Index) from 1992 to 2007

Year	*CPI*	*Year*	*CPI*	*Year*	*CPI*	*Year*	*CPI*
1992	106.4	1996	108.3	2000	100.4	2004	103.9
1993	114.7	1997	102.8	2001	100.7	2005	101.8
1994	124.1	1998	99.2	2002	99.2	2006	101.5
1995	117.1	1999	98.6	2003	101.2	2007	104.8

Table 9.11 The Degree of Technology Transfer in China from 1992 to 2004

Year	*1992*	*1993*	*1994*	*1995*	*1996*	*1997*
MTF	0.198	0.219	0.253	0.249	0.201	0.212
Year	1998	1999	2000	2001	2002	2003
MTF	0.264	0.265	0.237	0.246	0.338	0.368

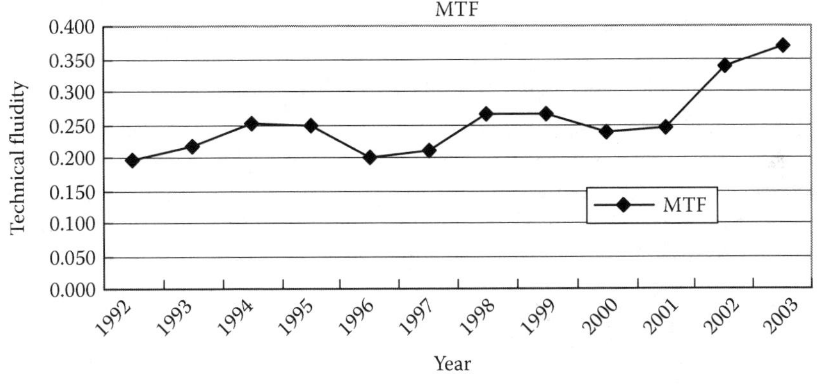

Figure 9.1 Line chart on the degree of China's technology transfer from 1992 to 2004.

b. The degree of technology transfer measurement

According to Formula 9.2 of China's measurement of the degree in technology transfer, the degree of technology transfer in China during the period 1992–2003 is shown in Table 9.11.

To reflect more directly on the development and changes in the degree of technology transfer, the line chart of technology transfer degree can be described as in Figure 9.1. As can be seen from the chart, China's technology transfer degree shows a rising trend of fluctuation in the process. The degree of technology transfer declines in 1996, 1999, 2000, 2003, and 2004, but the degree of transfer in other years shows a rising trend.

9.2 Empirical Analyses in Technology Transfer and Economic Growth in China

9.2.1 Technology Transfer and Economic Growth Theory

Adam Smith was the first man to find an expression for the impact of technology progress on economic growth in microeconomics. In his *The Wealth of Nations* he adopted the method of element analysis on economic growth and pointed out that labor, capital, and plot scale decide the total output of a nation and are essential elements for economic growth. He indicates that any annual yields of land and labor only increase in two ways, namely, improving effective productivity employed and increasing effective employment in societies. The improvement on effective productivity depends on the improvement of the ability of workers as well as their use of machines, all of which are related to technological advances. Smith had realized that technological progress is the potentially profound one besides the three basic elements in economic growth.

Schumpeter in his *Economic Development Theory* published in 1912 put forward the idea that rather than capital and labor, it is the technological innovation that is the main source of capitalistic economic growth. Schumpeter held that the capitalistic economic development is a process of dynamic balance which is highly affected by technology innovations. Furthermore, he stressed that as an exogenous economic variable, technology has a huge impact on economic growth. In his point of view, because of the motivation of reaping excessive profits, enterprises attach great importance to research and development activities. As technological progress and development of enterprises show a positive correlation, companies cannot develop without innovation.

Technological determinism scholars in the neoclassical theory of economic growth argue that technological progress is a major source of economic growth. U.S. economist R. Solow initially recognized that technological innovation is the main source of economic growth. He classified economic growth as from two different sources: one is the "growth effects" due to the increase in number of factors and the other is the economic growth of "level effect" due to improvement on the level of factors. The latter means that without increasing investment, technological progress can change the production function by moving upward without input of elements and consequently achieve the goal of economic growth.

Scholars with the new growth theory claim that technology plays a unique role in economic growth. Since the globalization in economy speeds up the transfer of technology in the world, Agawa illustrated the relationship between economic and technology transfer by analyzing the shift of the economic growth centre in the world, as shown in Table 9.12.

Owing to the global flow of technology the birthplace of techniques or industries are often unable to maintain its surplus at the end. From the table above, it is clear that the history of the world's technology transfer can be described on the follows lines: the Middle East—Southern Europe—Northern Europe—Britain—the Eastern United States—the United States and the Western Pacific region. At the same time, the global economic growth center of location is not far behind the birthplace of the technology industry.

In the Economics literature, the North–South trade in the technology transfer, imitation, innovation of the national economic development, and opening up has long been the main components of economic research.

Since the 1980s, Krugman who represented a group of economists created the "new trade theory" and the rapidly developed "product cycle theory" opening a new path for the study of technology transfer. The research on the product life cycle

Table 9.12 Economic Growth Center in the World

Geographical Center	*Geographical Centers of Economic Surplus Each Time*		
Technology/Industry	*Birthplace*	*Region of Economic Surplus*	*Time*
Sailing	Middle East	South Europe	From mid thirteenth to mid sixteenth century
Trade and banking	South Europe	North Europe	From mid sixteenth to mid eighteenth century
Management and control	North Europe	Great Britain	From mid eighteenth to mid nineteenth century
Industrial revolution	Britain	East United States	From early twentieth to late twentieth century
Information revolution	East United States.	The west and pacific region of United States.	From the end of twentieth century to twenty first century

by Vernon is a pioneer theory of the North–South trade and technology transfer. Krugman developed the proposition suggested by Vernon, the proposition became an official and general equilibrium model.

R. van Elkan (1996) established the general equilibrium model of technology transfer, imitation, and innovation under the condition of the open economy. In his view, the South and the North's economic development is convergent, that is countries with less developed economies can imitate, bring in, and innovate technology to achieve the economic catch up. The assumption that all countries' stock of human capital can be upgraded effectively through technology transfer, imitation, and innovation, of which the increase in potential productivity efficiency brought about by technology imitation depends on the technical differences between two countries. And the effectiveness of technology innovation depends on the accumulation of experience in "learning by doing." Through the effect of technology transfer and diffusion, any country's technology investments are likely to increase the domestic economy and income levels, as well as those in other countries.

Barro and Sala-I-Martin (1997) promoted low-cost imitation models, and demonstrated the economic systems that can be derived from the convergence of the nature of economic growth. The theory assumes that technology in different economic systems can diffuse, because the cost of copying is less than the cost of innovation. And those who follow economic growth will be faster than the technology leader, and the leader in technology and those who follow economic growth will at be the same rate of growth in the end-time state. Neoclassical growth theory predicted the convergence of global economic growth and that it will bring international technology transfer to less developed countries, developing countries the opportunity to catch up. Through their faster accumulation of capital and introduce advanced technology from developed countries, they can continue to narrow the gap between them and the developed countries.

In China's Center for Economic Research, Lin Yifu, in the view that if in a developing country the government's development strategy deviated from the optimal choice of technology, it will have an impact on the country's economic growth rate. It will impact the outcome whether the country's economic growth and that of the developed countries is able to converge to the income level or not. Fu Qiang, who discussed the impact of industrial structure and technology transfer on the economic growth model, pointed out that if FDI is accompanied by technology transfer and service, it may delay the introduction of technology in the country's economic growth. In addition, more Chinese scholars analyze the combination of multinational investment in China to conduct the technology transfer to China to productivity related industries, technology research, development capabilities, labor employment, and so on. Wu Lin (2002) investigates Suzhou, Wuxi, Changzhou, three-level science and technology park, and analyzes the role of transnational corporations in China's strategy for technology transfer and science and that of the technology park on the spillover effects.

9.2.2 Demonstration Test

1. Granger causality test model (the part: Liupu Li master's thesis)
   a. The stationarity test of time series

To test the Granger causality of the technology transfer and economic growth, the general practice, based on a sample of the existing data, is to establish the more appropriate regression equation. In the traditional regression analysis, the time series requires to be stable, or there will be "Spurious Regression." However, in reality, the economic time series are often nonsteady, which destroys the assumption of the steady destruction. To make the return meaningful, time series need to be steady, and the common practice is to put difference into time series, and then use the difference sequence to return.

In general, if the average and variance of the time series maintain a constant mean and variance at any time, the covariance (or self-covariance) between the two periods t and $t+k$ only depends on the distance k between the two periods, nothing to do with the actual period of the calculation of these covariance, then the time series is steady. Otherwise, as long as it does not fully meet the three conditions, the time series is unsteady.

Unit root is a method for the nonstationary series. Unit root method will make the test of nonstationary series into the test of the unit root. If the variable x_i has unit root, the process of testing the stability of the variables is called the unit root test. If the variables have a steady process, it is expressed as $I(0)$; if it becomes a steady process after the first difference, it is known as the unit root process, which is expressed as $I(1)$. We use the augmented Dickey–Fuller (ADF) unit root test to test, and the test regression equation is as follows:

$$\Delta x_t = \alpha_0 + \alpha_1 t + \alpha_2 x_{t-1} + \sum_{i=1}^{k} \alpha_{3i} \Delta x_{t-i} + \mu_t \tag{9.3}$$

where

α_0 stands for the constant
t stands for the time trend
μ_t stands for the residual item

The original hypothesis is $H_0 : \alpha_1 = 0$, and the alternative hypothesis is $H_1 : \alpha_1 < 0$. According to the value of the t test of the coefficient α_1, in the regression equation, if the test t value is less than the critical value of the distribution of ADF, the original assumption is rejected and the alternative hypothesis accepted, which indicates that the sequence $\{x_t\}$ is a steady process. If the test t value is greater than the threshold, the original assumption that the sequence unit root exists is accepted. To make the residual difference μ_t become the white noise, the equation is added to seven of the lagged variables.

b. Granger causality test

Generally speaking, to carry out the Granger causality test, the time series must be proved steady only through the unit root test. In the regression analysis, regression can measure the link between the variables, but it cannot prove cause and effect. Identifying the cause and effect is an important issue in the study which is based on the test. The basic idea of the causality test brought forward by Sim and Granger is as follows. If the variables X contribute to the forecast variables Y, that is, the reunification of Y is carried out according to the Y value in the past. If coupled with the X value in the past, which can significantly enhance the explanation ability of regression, X is a Granger cause of Y, others are called the causes of non-Granger. The course of the Granger causality test between variables X and Y is as follows:

$$y_t = \alpha_0 + \sum_{i=1}^{n} \alpha_i y_{t-i} + \sum_{i=1}^{n} \beta_i x_{t-i} + \mu_t \tag{9.4}$$

α_0 is constant, lag phase n is the arbitrary choice, testing X is not quite the Granger causes of Y is equal to the test of F of the statistical original assumptions $H_0 : \beta_1 = \beta_2 = \ldots = \beta_n = 0$. RSS_1 stands for the regression of the residual sum of squares, and RSS_0 stands for the sum of residual regression squares. When the original assumptions come into existence, the test statistic is as follows:

$$F = \frac{(RSS_0 - RSS_1)/n}{RSS_1/(N - 2n - 1)} \tag{9.5}$$

N is the sample size. F is the test statistic subordinate to the standard F distribution, if the F test value is greater than the threshold of the standard F distribution, the original assumption is rejected, which shows the change in X is the Granger reasons of Y, otherwise accept the original assumption that the change in X is not the Granger reasons of Y. By the same token, if the location of X is exchanged with Y in the regression equation, Y can be used to determine whether it is the Granger reason of X.

2. Demonstration analysis
   a. The steady test of time series

According to Table 9.13, the unit root test of MTF shows that the ADF statistics is minus 4.477163, less than 5 percent of the significant level of the critical value minus 3.9948, so the original assumption is rejected. MTF has no unit root, that is, the MTF is a steady variable.

Table 9.13 Test Table for the Stationarity in the Degree of Technology Transfer

ADF test statistic	–4.477163	1 percent critical value	–5.2735
		5 percent critical value	–3.9948
		10 percent critical value	–3.4455

Table 9.14 The Stationarity Test Table of GDP Growth Rate

ADF test statistic	–2.634784	1 percent critical value	–4.2207
		5 percent critical value	–3.1801
		10 percent critical value	–2.5349

Table 9.15 Technology Transfer and Economic Growth Causality Test Table

Null Hypothesis:	*Observations*	*F-Statistic*	*Probability*
MTF does not Granger cause GDP	11	2.72866	0.09117
GDP does not Granger cause MTF		3.61062	0.04394

According to Table 9.14, GDP growth rate of the unit root test shows that the ADF statistics is minus 2.635, less than 10 percent of the significant level of the critical value minus 2.535, so the original assumption is rejected. It means that the GDP growth rate variable has no unit root, that is, GDP growth rate is a steady variable.

In a word, through unit root test, the degree of technology transfer and GDP growth rate are the steady variables, so spurious regressions do not exist when the causality between the degree of technology transfer and the GDP growth rate is analyzed. Hence, the causality between the degree of technology transfer and economic growth analysis is analyzed.

b. Analysis of causality in the degree of technology transfer and economic growth

Statistical data Eviews 5.0 software is used to analyze causality between the degree of technology transfer and economic growth data. After testing, at the significance level of 5 percent, GDP growth is the Granger cause of the degree of technology transfer growth; at the significant level of 10 percent, the degree of technology transfer is also the Granger cause of GDP growth, which are seen in Table 9.15. As a result, we can say that at the 10 percent level of significance, there is a two-way causal relationship between Chinese technology transfer degree and economic growth. The technology transfer promotes economic growth, while economic growth has also brought about technology transfer.

9.3 Technology Spillover Effects of FDI and Its Relationship with the Growth of Chinese Economy

China attracts the most FDIs among developing countries; the main purpose is to introduce the advanced technology of the foreigners and impose the advanced technology to flow into the country, with the dominant aim to supplement the fund. The direct investment of foreigners has already become one of the most important ways to introduce the advanced technology from those nations.

9.3.1 Characteristics of FDI Introduced by China

Under the surrounding downcast world economic scene and the decreasing investment of transnational corporations, China has continued to keep a high level in attracting the direct investment of foreigners and the use of foreign funds has been the top among the developing countries and areas for 12 years now.

China had the highest FDI among developing countries in 1993 and the industries department was the top in all the branches, and even exceeded USA to become the No. 1 in 2002. From the data of the Statistics Bureau of China, the number of contracts that use the investment of foreigners has reached 43,664, which increased by 6.3 percent compared to 41,081 of the last year and only 34,171 in 2002. The funds invested by the foreigners in contracts has increased rapidly to $1,534.79 hundred million which is an increase of 33.38 percent compared to 2003 and the actual use of foreign investment is $606.30 hundred million, increased by 13.32 percent. The introduction of foreign investment has been kept at a perfect level in China. By 2005, China had already sanctioned 508,941 projects with investment by foreigners, the fund with investment by foreigners in contracts had reached 10,966.09 hundred million, foreign investment of 5621.05 hundred million was actually used.

The entry of foreign investment has directly injected capital into the economy of China, which ensures forcefully to the increase of economical output; on the other hand, the way China has introduced the investment can give an increase to citizen welfare (mainly in the benefit of outflow of technology), besides the scope of economic development. It can give a strongly dynamic push to the economic level by connotation.

The situation created by direct investment of foreign businesses in China has the following characteristics:

1. Its scale has increased at a steady pace. From 1985, the fund introduced in the foreign country has improved without decrease has only seen some fluctuation in the speed between the years.
2. It has the domain position in the economics of China, which is different compared to many developing countries in this aspect. The way the foreign fund is used is mainly in introducing direct foreign investment, and the proportion of foreign investment is increasing continuously. In 1985, the actual total use of foreign investment in China was $47.60 hundred million, the directly investment of the foreign business was $19.56 hundred million and accounted for 41.09 percent of the total investment. In 2004, this proportion has increased to 75.40 percent.
3. It is distributed extensively in real estate and spares in China and industries that concentrate on spares attract a lot of investment and the east area is the main place that has foreign investment. After a long period of reform and the opening up policy, the foreign investment commenced in the south eastern coastal provinces, but in recent years the foreign business has moved to the middle and west at a high speed. From the areas the foreign businesses invest in, it is clear, that at the beginning of the reform and opening up policy, it focused on the machine labor-concentrated industries, then the investments went into the fund-concentrated industries; the large scale of investment in the car industry is an apparent example.
4. Its inner quality takes on a step by step increasing tendency, which is evident from two aspects: first, there are more and more transnational corporations, such as the top 500 companies in the world, that chose to invest in China this year, an obviously contrast to the beginning of the reform and open policy, when only small and middle companies invested in China; second, the transnational corporations have carried out their research and action into China. The expenditure of R&D action of FDI businesses has increased rapidly.

9.3.2 Analysis of the Technology Spillover Effects in FDI

According to the literature, FDI flows into the host country result in the realization of the following effects.

1. Demonstration–imitation effect

Demonstration–imitation effect is the contagion effect discussed by Kokko. That is, because of the technology gap between the FDI enterprises and enterprises of the host country, the host country enterprises may learn to imitate their behavior to improve their productivity and skills. Foreign-funded enterprises will not only introduce new equipment, new products, or new processing methods into the domestic market, but also bring into the product selection, marketing strategy, management philosophy, and other nonphysical and chemical techniques. In some cases, domestic firms can improve their productivity only through observing and learning from the nearby foreign-funded companies.

A questionnaire survey made by Stephen Young and Ping Lan, the Chinese managers of the Dalian joint venture enterprise, showed that 26 percent of people felt that high-level access to the technology was achieved by this (which may help shorten the technology gap by more than ten years), and 48 percent felt a mid-level in technology transfer was achieved.

2. Competitive effects

The competitive effect often occurs between vendors in an industry. The effects of competition are twofold—on the one hand, they encourage scrabbling by the foreign-funded enterprises and enterprises of the host country for the limited resources of the market, which increases competition in the market and stimulates more efficient use of existing resources by local manufacturers to promote local technological efficiency; on the other hand, they raise the level of social welfare, because the foreign-funded enterprises forcing their way into the industry against strong industry barriers, eliminates monopolies to a certain extent.

Wang and Blomstrom built the game's basic business model based on the transnational corporations and the local subsidiaries to prove that competition promoted the local enterprise technology progress, and narrowed the technology gap between the two, and the transnational companies were forced to introduce or develop new technology to maintain their comparative advantage in technology, which also led to a new round of technology spillover.

3. Contact effect

Contact effect is seen as the flow links in industrial technology, where foreign-invested enterprises communicate with the local business contacts or customers, including the backward link to the upper reaches of enterprise suppliers and the forward link to the lower reaches of enterprises such as the vendors. Yifu Lin, Xinqiao Ping, and Dayong Yang completed a case study which illustrates the effect of foreign capital industry contact. They calculated the impact of the Coca-Cola enterprise on the overall Chinese economy. The production in China of the Coca-Cola cans not only brought benefits to the production workers directly, but also affected the lower reaches of the sales and increased the final demand, showing that all direct and indirect effects need to be taken into account. They calculated the overall effect through the Chinese "input–output table" in 1992. Research shows that the Chinese economic output multiplier was impacted about 2.66 times by the Coca-Cola canning enterprise in 1998, and its upstream and downstream economic activity have created about 400,000 jobs.

W. Mark Fruin and Penelope Prime (1999) compared the United States-funded enterprises with the Japanese-funded enterprises in the Chinese competitive strategy, and pointed out that Japanese-funded enterprises required a high level of operational efficiency, and focused on the supply of training and quality control of products. For example, suppliers improve product quality and promote innovative activities for technical assistance and provide information services, as well as organizational management training and help. All these are constitute the contact effect leading to the spillover effect.

4. Training effect

The experience of developed countries confirms that foreign capital has a competitive advantage that cannot be divorced from its human resources and is fully materialized in equipment and technology. As a result, the effective functioning of the transnational companies' overseas investment projects often go hand in hand with local human resource development. This involves the local technical and management staff working with the experts from the headquarters of the transnational companies, training of the local personnel, participation of the local personnel in the research and development activities for the technology, product, and process improvements, and senior managers learning and taking part in the process of the multinational companies and so on.

In the Hong Kong technology transfer study, Chen found that the affected areas and expenditure of the transnational corporations times those of the local businesses in the three quarters of the sample enterprises. He concluded that the largest contribution of foreign enterprises did not lie in the so-called new techniques and products, but in the training of workers at all levels in the Hong Kong manufacturing sector. When employees of a subsidiary of a transnational corporation move to local companies or resign to create their own businesses, the skills and experience learned when they were employed with the foreign enterprise are transferred to the new enterprise or business. This is an example of the FDI training effect leading to technology spillover caused by the loss of human capital in the foreign enterprise.

5. Aggregation effect

Aggregation effect usually refers to the positive external economic effects resulting from a number of specific economic activities. There has been a lot of research on the aggregation effect in China and abroad, but analysis of FDI technology transfer is not much. In fact, phenomenon, such as concentration of FDI in cities, in the same industry or space, or in the same country in one location can be found everywhere in developing countries and the aggregation effect plays an important role in the process of FDI for the spillover in the host country.

A large number of studies have shown that accumulation economy impacts positively on the locations where foreign economic manufacturing plants are located.

The accumulation economy is reflected not only on the level of city economies, but also on that of industries. A large number of related businesses gather together in a region, which can save production costs, expand production and consumption demand, is conducive to competition and collaboration, and improvement in the management and work performance.

Feng Wu, referring to the choice of location in specific countries from the point of view of FDI, pointed out that the gathering effect of FDI is an important factor in site selection, and that the fractionized aggregation effect can, on a theoretical level analysis, be carried into five-level effects. The five-level effects are the gathering effects of urbanization, of specific industries, of foreign investment, of particular investment in a country or region, and of specific companies. From the host country's point of view, the gathering of FDI means more spillover. First, it will inevitably lead to competition and promote technology innovation, the concentration itself makes for the diffusion and spreading of the innovation and the advantages of technology and the market brought about by innovation further promotes the investment of companies in production, which constitutes investment in R&D. Second, the mutual competition or cooperation of the small and medium enterprises forms the dynamic enterprise network, producing an external economy; therefore the small and medium enterprises can absorb the advanced technology transferred by the parent company of transnational companies. Once the host country enterprises participate in this network, they also enjoy the specialized division of labor, resource sharing, transaction costs saving, the close link of input and output between manufacturers, and other benefits.

9.3.3 Empirical Test of FDI Technology Spillover Effects

An analysis of FDI technology spillover effects, raises the question whether FDI in China has all the aspects of the technology spillover effects, or has brought economic efficiency of the future. The country's total annual data can be used to test the Chinese FDI technology spillover effects by selecting the appropriate model.

1. Spillover model
   a. The choice of variables interpreted

The variable best suited to reflect Chinese technology progress indicators is used. Despite a number of measures of economic growth of technical factors, the ideal way is to get a specific form of the production function, add elements that affect the independence of the problem, and so determine the total factor productivity index which will indicate the technology progress from the C–D function. Determination of total factor productivity index with specific methods is as follows:
Assuming C–D function is as follows:

$$Y = AL^{\alpha}K^{1-\alpha} \tag{9.6}$$

where

A stands for the technology level
L stands for the labor inputs
K stands for the capital
Y stands for the total output level
α stands for the output flexibility of the labor output
1 – α stands for the output flexibility of capital

the rate of technology progress can be expressed as

$$A_t/A_0 = \frac{Y_t}{L_t^{\alpha}K_t^{1-\alpha}} \Big/ \frac{Y_0}{L_0^{\alpha}K_0^{1-\alpha}} \tag{9.7}$$

As a result, the total factor productivity index is defined as

$$F_t = A_t/A_0 = \frac{Y_t}{L_t^{\alpha}K_t^{1-\alpha}} \Big/ \frac{Y_0}{L_0^{\alpha}K_0^{1-\alpha}} = \frac{Y_t/Y_0}{(L_t/L_0)^{\alpha}(K_t/K_0)^{1-\alpha}} \tag{9.8}$$

b. The explanatory variables and the model choice

FDI and R&D investment are chosen to explain variables, and the model adopted is as follows:

$$\text{Ln}F_t = a + b\text{LnFDI}_t + c\text{LnR\&D}_t + \varepsilon_t \tag{9.9}$$

$$\text{Ln}F_t = a + b\text{LnFDI}_t + b_{-1}\text{LnFDI}_{t-1} + c\text{LnR\&D}_t + c_{-1}\text{LnR\&D}_{t-1} + \varepsilon_t \tag{9.10}$$

The main difference between models (9.10) and (9.9) is in the lag phase of FDI technology spillover effect on the Chinese technology progress, which is the effect of the total factor productivity.

c. Model target and data specification

First, it should be noted that China has statistical data on R&D expenditure from the beginning of 1995, and prior to 1995 there is no related data. So, R&D expenditure will not be included in the test model. To solve this problem, the R&D expenditure data trend curve was observed and it was found that the distribution is an exponential distribution, so an index function was constructed to forecast the R&D expenditure data from 1990 to 1994, and the data can be seen in Table 9.17.

From its contribution to the output, R&D expenditures on output will certainly confirm the existence of a long period of lagging behind, because their R&D activities have the characteristics of a longer cycle and their technology achievements transfer into actual use of capital and skills only after a period of time. Model 9.10 is more reasonable in the theory behind the introduction of R&D, but the actual test is also needed.

Models (9.10) and (9.9) do not include the effect of quality human capital on total factor productivity. In fact, if the models (9.10) and (9.9) include the use of human capital factors, they can reflect more accurately the total factor productivity, and measure the FDI spillover effects accurately. However, it is difficult to use this model to test empirically, primarily because data comparison on labor quality is difficult to obtain.

In Formula 9.8 of the total factor productivity index calculation, the determination of capital K has been very controversial in the academic community; some scholars suggest using capital stock, others suggest using capital flows. After analysis, it seems more appropriate for capital stock data to represent capital K. The current output Y in the production function is not only the result of the current capital input K and the labor force element L, but also has a close relation to the preliminary capital investment. It can be said that capital formation together with the original capital deposit contributes to the output of each later period, so the capital of the production function adopted as the current fixed capital would be wrong and capital stock data would be more appropriate to represent capital K.

In the *Statistical Yearbook of China*, capital stock data cannot be found. Generally, the academic community adopts the so-called perpetual inventory to determine the capital stock data. These general steps are as follows: first, through the census or according to some estimates, the calculation of the capital stock of the whole society during a base period is assumed; and then according to the principle of the capital stock increasing in the previous year and the incremental capital of the current year equivalent to the capital stock deposit of the current year, capital stock deposit data of the calendar year is estimated, and then the constant prices of the calendar year are amended according to the constant prices in 1990. The specific data is seen in Table 9.16.

In this chapter, employment is in place of labor input L at the end of the Chinese calendar year; assuming that there is no qualitative difference in the labor force between individual Chinese workers. The difference in the labor force between individual workers is reflected by the input elements of the progress of science and technology, because the improvement in the quality of workers is the result of the technological and scientific progress. The labor data is given in Table 9.16.

The National Development and Reform Commission suggested that the output of the proposed funding flexibility was 0.35, and that for flexibility in labor output was 0.65 estimating the contribution rate of science and technology industry. This suggestion was adopted to measure the total factor productivity index, and α was taken as 0.35 and β as 0.65. In theory, α and β should be variables; with

Table 9.16 Total Factor Productivity Index Measurement Data and the Results of Calculations

Year	*The Total Output (Y: Hundred Million ¥)*	*Capital Input (K: Hundred Million ¥)*	*Labor Input (L: Ten Thousand Persons)*	*Total Factor Productivity Index (F)*
1990	18547.9	64,850	64,749	1
1991	20253.2	70,091	65,491	1.055
1992	23137.1	77,109	66,152	1.158
1993	26258.1	87,020	66,808	1.252
1994	29583.1	97,802	67,455	1.345
1995	32690.9	108,993	68,065	1.423
1996	35825	121,085	68,950	1.490
1997	38992.1	134,145	69,802	1.552
1998	42040.6	149,388	70,637	1.599
1999	45043	165,774	71,394	1.641
2000	48644.6	183,672	72,085	1.699
2001	52292.3	203,669	73,025	1.747
2002	56631.3	227,092	73,740	1.809
2003	61898.8	256,426	74,432	1.884
2004	67903.6	291,390	75,200	1.963

Source: National Bureau of Statistics of China, *Chinese Statistical Yearbook, (2005)*. China Statistical Press, Beijing, IO, 2006.

Note: The total factor productivity index is for the set-based index.

the improvement of scientific and technological progress, the output flexibility of capital and labor should be increased, but the increase should be brought about by scientific and technological progress, so the fixed values for α, β could estimate the total factor productivity more comprehensively.

Y is the GDP expressed by the fixed price; it is an index expressed in currency in keeping with the current year's prices. The factors that affect the price must be eliminated to reflect factually the economic development dynamics, because it contains the annual price changes and hence cannot exactly reflect the increase or decrease of the amount of physical changes in the comparison between different years. In this chapter, the GDP is seen as a measure of the total output of the basic indicators on the basis of data from the *China Statistical Yearbook* (2005), and based on the 1990 constant prices and the index for discount, the specific formula is as follows:

$$\text{GDP}(1990 \text{ fixed price}) = (\text{GDP index} / \text{GDP index in} 1990) \times (\text{GDP in } 1990) \tag{9.11}$$

2. Empirical test of FDI technology spillover effect
   a. Total factor productivity measurement

According to Formula 9.8 the total factor productivity index calculation and a number of processing methods of the above-mentioned indicator data, calculations, and the data used by the results of the calculation can be seen in Table 9.16.

To describe more clearly the Chinese rate of technology progress, the estimated total factor productivity index is plotted as a curve in Figure 9.2. As can be seen from the chart 9.3, the technical level in China is increasing year by year. There are two characteristics of the rate of technology progress measured by us: first, it fits in well with the results of experience; second, many scholars use the Solow residual value to estimate the rate of technological progress. But sometimes the rate of technological progress is negative, which means that the technical level slides down. Between the years 1990–2004 it is difficult to explain the negative values; the rate of technology progress measured is positive instead of negative, which indicates the level of technology is on the rise.

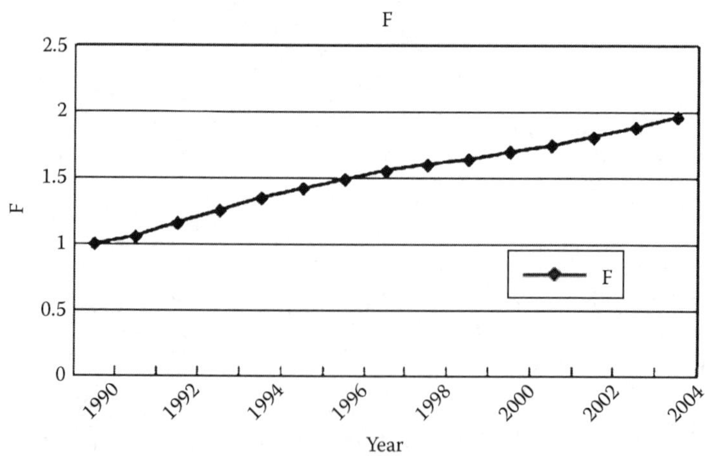

Figure 9.2 Chinese technological progress rate curve.

b. Empirical test of FDI technology spillover effect

In the models (9.9) and (9.10), empirical research is designed to study the contribution of FDI to the total factor productivity, which is the criterion for the technology spillover effect of FDI. To test the technology spillover effect of FDI, a different interpretation of variables has been established, including for the four models of FDI for the lagged first phase and lagged second phase and the effects for the total factor productivity have been explored, as well as the introduction of R&D lag factor to analyze these changes in the above impact.

FDI spillover effects of empirical test data are used as a list in Table 9.17.

The test results of FDI spillover effects can be seen in Table 9.18. The results of the four regression equations show that the overall impact of FDI total factor productivity is positive, and is significant. Derived from the coefficient of determination, it is clear that the four models fit very well. The parameters of the T-test also indicate that the estimated parameters of the variables are significantly different from zero.

Model I carried out a regression to the current period's value of FDI and the current period's value of R&D. It is clear that the various parameters of the T-test values are very significant and the models fit better, but in the four models it fits badly, which indicates that FDI and R&D technology spillover exist as factors that lag time.

Model II carried out a regression to the current value of FDI, the lagged first phase of the value of FDI, and the current value of R&D. The results show that the FDI modulus in the current period is significant, which indicates that FDI on the impact of total factor productivity does need a period of time before gradually playing out. In combination with the technical factors of FDI, it appears that the contribution of the total factor on the FDI growth rate is because it takes some time for local companies to absorb these techniques. This result fits well with the above analysis of the FDI technology spillover effects.

Model III carried out a regression of the current value of FDI, the lagged first phase of the value of R&D, which fits better compared with Model I and indicates that R&D has the same impact as FDI on total factor productivity, which needs a period of time to play out gradually.

Model IV can better describe the problem. This model introduces FDI and the delayed second phase of the value of R&D after the value of the second phase, the results show that this model fits best and its parameters of the T-test are significant, which is larger than the previous three models. Lagged FDI variable greatly affects the total factor productivity (The greater the coefficient parameter is, the higher the confidence level is), which indicates that FDI has a longer time-lag technology spillover.

In a word, the results of the study shows that FDI technology spillover effect is significant, and the spillover delay effects are apparent.

Table 9.17 FDI Spillover Effects of Empirical Test Data

Year	*FDI (Hundred Million $)*	*Amendments to FDI (Million)*	*LNFDI*	*R&D (Hundred Million ¥)*	*Amendments to R&D (Hundred Million ¥)*	*LNR&D*
1990	34.87	166.79	5.12	127.37	127.37	4.85
1991	43.66	217.74	5.38	154.39	144.65	4.97
1992	110.07	527.22	6.27	187.16	162.56	5.09
1993	275.15	1201.98	7.09	226.87	172.00	5.15
1994	337.67	1841.24	7.52	275.02	173.99	5.16
1995	375.21	1751.65	7.47	348.69	194.93	5.27
1996	417.26	1830.81	7.51	404.48	213.46	5.36
1997	452.57	1964.57	7.58	509.16	266.62	5.59
1998	454.63	2019.75	7.61	551.12	295.73	5.69
1999	403.19	1831.92	7.51	678.91	372.62	5.92
2000	407.15	1832.60	7.51	895.66	486.98	6.19
2001	468.78	2084.98	7.64	1042.49	560.18	6.33
2002	527.43	2350.68	7.76	1287.64	693.35	6.54
2003	535.05	2339.39	7.76	1539.63	813.30	6.70
2004	606.3	2489.52	7.82	1966.33	975.49	6.88

Note: The amendment to the FDI is that the FDI data will be used to reduce GDP revised index values; the amendment to the R&D was amended similar to the amendment to FDI.

Table 9.18 Empirical Results of the Four FDI Technology Spillover Effects Models

	Return Data							
	Model I		*Model II*		*Model III*		*Model IV*	
Variable	*A*	*Pr* > |*t*|	*A*	*Pr* > |*t*|	*A*	*Pr* > |*t*|	*A*	*Pr* > |*t*|
Constant	1.194	<.0001	–1.033	<.0001	–1.3008	<.0001	–1.238	0.0004
LNFDI	0.129	<.0001	0.018	0.5112	0.1373	0.0002	0.158	0.0800
LNFDI(–1)			0.108	0.0006			0.152	0.0704
LNFDI(–2)							0.160	0.0009
LNR&D	0.310	<.0001	0.289	<.0001				
LNR&D(–1)					0.3267	<.0001		
LNR&D(–2)							0.2916	<.0001
R^2	0.9479		0.9659		0.9852		0.9960	

Note: A stands for the parameter values.

9.3.4 FDI and Its Technology Spillover Effects and China's Economic Growth

Economic growth has always been one of the most important issues of economics, so to understand economic growth is very important. As a modern pioneer of the theory of economic growth, Adam Smith, in his paper called "The nature and causes of wealth of the nation" proposed that the engine of growth lies in the division of labor, capital accumulation, and technology progress. The theory of economic growth has been closely connected to the study of the technology progress, from exogenous technology progress of Harold, Solow, and Denison, innovation-driven technology progress of Schumpeter Kuznets to the neoclassical endogenous growth theory of technology progress represented by Romer and Lucas. These theories have emphasized the growth of technological progress as the ultimate contribution to economic growth.

Through the empirical test of the FDI technology spillover effect, it was found that there was a significant relevance between the FDI flows and China's technology progress, that is, total factor productivity and FDI flows play a very important role in China's technology progress. Therefore, according to the theory of economic growth, a large number of FDI flows will bring up the China's economic growth. The following is the empirical analysis of the relationship between FDI and Chinese economic growth.

1. Determination of the relationship between the GDP growth and FDI

From 1983 to 2004, China has made use of FDI which rose from $ 636,000,000 in 1983 to $ 62,700,000,000 in 2004, an increase of 97.6 times during these 22 years. Among the developing countries China has been able to attract the largest foreign investment in the last consecutive 17 years, and was ranked second in the world after the United States in the use of foreign capital. The FDI in 1983, 1992, and 1993 was 6.36, 110.07, and 275.15 hundred million $ respectively and the growth rate of FDI for these three years was the fastest, and was 97.80 percent, 152.11 percent, and 149.98 percent respectively over the previous year's growth rate. Because of the Asian financial crisis in 1997, FDI in 1998 grew by only 0.45 percent, and the first negative growth appeared in 1999. China overcame the initial impact of the Asian financial crisis in 2000, and the recovery signs of FDI were clear. Analysis shows that FDI growth is similar to GDP growth: FDI increases rapidly when the economic situation is better, such as in the years from 1992 to 1993, its growth rate has a corresponding slowdown or even negative growth when the economic situation deteriorates, such as in the years from 1988 to 1989 and from 1997 to 1999. In particular, prior to 1994 (excluding the years from 1988 to 1989 due to the impact of political instability on the economy), the absolute amount of the China's FDI growth is rising; but after 1994, China's FDI growth rate slowed down significantly. It can be inferred that FDI is interrelated to GDP. GDP growth is relatively fast in the year of rapid growth of FDI; whereas GDP growth is relatively slow in the year of slower growth in FDI.

2. Cointegration analysis between FDI and GDP

Cointegration analysis is a new econometric result since the 1980s. The Engle–Granger two-step method or the cointegration system testing method based on VAR methods could be used to test the long-term balanced relationship between two nonstationary time series.

a. Data and model

The following two variables GDP and FDI are considered for this analysis. In the empirical analysis of data, the years from 1983 to 2004 are taken as a sample interval and the data from *Chinese Foreign Trade and Economic Yearbook* and *China Statistical Yearbook* of the previous years is used. To facilitate the study, a steady sequence is generated by taking the logarithm of the data sequence number, which does not change the characteristics of the variables. The logarithm of the variables of GDP, FDI are taken to get new variable sequences recorded as LNGDP and LNFDI.
The general regression model is as follows:

$$\text{LnGDP} = C(1) + C(2) * \text{LnFDI} \tag{9.12}$$

b. Unit root test

As most of the economy time series are nonstationary, cointegration tests must be carried out by the unit root test, only if the variables are the same as in the single-order, then the whole sequence can be used in the cointegration regression. Before using the method, the entire single-whole tests must be done after the timing variables are analyzed first. If a sequence becomes a steady sequence after the difference *d* times, then the sequence is called *d*-order single whole, and marked as I(d). One-time inspection of the entire order is whether it is $I(0)$, and then whether it is $I(1)$, the discrimination is based on the ADF test of unit root test. The ADF test is used to test the stability of variables, such as test results in Table 9.19.

Table 9.19 shows that all the level series of the variables are not smooth, and their first-order differentials are stable, that is, all are $I(1)$ sequences; the two long-term relationship between the variables can be established through the next step of cointegration test.

3. Cointegration test

In their papers, Johansen (1998) and Johansen-Juselius (1990) put forward the cointegration method based on the VAR methods to test the system. According to the pricing guidelines formula by AIC:

Table 9.19 Unit Root Test Results of GDP and FDI

Variable	*Test in the Form of (c, t, and k)*	*ADF Value*	*5 Percent Threshold*	*Conclusion*
LNGDP	(c, 0, 1)	−0.37	−3.04	Not steady
|Δ|LNGDP	(c, 0, 1)	−4.03	−3.88	Steady
LNFDI	(c, 0, 1)	−0.93	−3.04	Not steady
ΔFDI	(c, t, 2)	−4.45	−3.73	Steady

Note: ADF test results are calculated by using Eviews software, and the forms of inspection (*c*, *t*, and *k*) unit root test respectively stand by unit root equation including the constant, the trend of the time and lag rank, and Δ is difference operator.

Table 9.20 Johansen Cointegration Test Results of FDI and GDP

Assuming the Original H_0	*Alternative Hypothesis H_1*	*Likelihood Ratio*	*5 Percent Threshold*	*1 Percent Threshold*
$r = 0$	$r = 1$	20.9	18.2	23.46
$r \leq 1$	$r = 2$	6.3	3.7	6.4

Note: The table test results are calculated by using EVIEWS software, and *r* stands for the number of representatives of the cointegration vector.

$$\text{AIC} = \text{Log}\left(\frac{\Sigma \varepsilon_i^2}{N} + \frac{2K}{N}\right) \tag{9.13}$$

where $\frac{\Sigma \varepsilon_i^2}{N}$ is the sum of residual squares ordinary least squares (OLS) of *Kth* order VAR models; the test results are in Table 9.20.

In Table 9.20, the results show that there is a unique cointegration relationship between the GDP and FDI variables at the critical level of 5 percent, that is to say there is long-term stability and balanced relations between the two. The corresponding cointegration equation is as follows including the intercept and trends variables:

$$\text{LNGDP} = -7.898 + 0.112 \cdot \text{LNFDI} - 0.123 \cdot \text{TREND} \tag{9.14}$$

As can be seen there is a positive correlation between FDI and GDP growth in China and the two have a relationship of long-term dependency from the cointegration regression equation 9.14. Test results show that there is a positive correlation between FDI and China's economic growth, that is to say the increase in FDI promotes economic growth. There is long-term stability instead of only short-term effect between GDP and FDI.

References

Aoki, R. and Tauman, Y. Patent licensing with spillovers. *Economics Letters*, October 2001, 73(1), 125–130.

D' Aspremont and Minde, J. Cooperative and Non-cooperative R&D in a Duopoly with Spillover. 1998.

Bao, J. The research of diffusion of technology innovation and its basic mathematical model. *Journal of Hefei Union University*, 2000, 10(3), 92–96.

Baptista, R. Geographical clusters and innovation diffusion. *Technological Forecasting and Social Change*, January 2001, 66(1), 31–46.

Barro, R.J. and Sala-i-Martin, X. Technological diffusion, convergence, and growth. *Journal of Economic Growth*, March 1997, 2(1), 1–26.

Batz, F.-J., Peters, K.J., and Janssen, W. The influence of technology characteristics on the rate and speed of adoption. *Agricultural Economics*, October 1999, 21(2), 121–130.

Bejean, L. Barriers to efficient monitoring of science, technology and innovation through public policy. *Science and Public Policy*, 1989, 16(6), 12, 345–352.

Bischi, G.-I., Dawid, H., and Kopel, M. Spillover effects and the evolution of firm clusters. *Journal of Economic Behavior and Organization*, January 2003, 50(1), 47–75.

Blomgren, S. Changes to building research funding in Sweden. *Building Research & Information*, 2003, 31(6), 479–484.

Boisot, M.H. *Information Space: A Framework for Learning in Organizations*, Institutions and Culture, Routledge, London, U.K., 1995.

Borenztein, E., Gregorio, J.D., and Leew, J.-W. How does foreign direct investment affect economic growth? *Journal of International Economics*, 1998, 45, 115–135.

Buzzacchi, L., Colombo, M. and Mariotti, S. Technological regimes and innovation in services: The case of the Italian banking industry. *Research Policy*, January 1995, 24(1), 151–168.

Cai, X. and Shi, H. Technical innovation and its positive research. *Science Research Management*, 1995, 16(6), 22–26.

Calderini, M. and Garrone, P. Liberlisation, industry turmoil and the balance of R&D activities. *Information Economics and Policy*, 2001, 13(2), 199–230.

Capper, C.A. An otherist poststructural perspective of the knowledge base in educational administration. In R. Donmoyer, M. Imber, and J.J. Scheurich (Eds.), *The Knowledge Base in Educational Administration: Multiple Perspectives*, Suny Press, Albany, NY, 1995, pp. 285–299.

Chen, X. Generation and influencing factors of technology spillover effect. *Fuzhou University Journal (Philosophy and Social Science Edition)*, 1999, 13(2), 23–27.

Chen, Q. and Lu, C. Study on the function innovation combined with diffusion stage of technology innovation. *Science of Science and Management of S & T*, 2001, 11, 47–50.

Chen, G. and Wang, X. New relationship argument among technology innovation, technology diffusion and technology advancement. *Studies in Science of Science*, 1995, 13(4), 68–73.

Chen, W., Dong, X., Sun, Y., and Geroski, P.A. Models of technology diffusion. *Research Policy*, April 2000, 29(4–5), 603–625.

Chen, Y., Shi, H., and Liu, S. Productivity flow model & demonstration research. *Statistics and Decision*, 2007, 21, 4–7.

Cheng, M. Study on general theory of technology innovation and diffusion process. *Economic Geography*, 1995, 15(2), 32–38.

Chu, X. and Liu, S. The investigation of liquidity premium based on Markov regime-switching model. *Systems Engineering*, 2007, 25(10), 16–20.

Colombo, M.G. and Mosconi, R. Complementarity and cumulative learning effects in the early diffusion of multiple technologies. *The Journal of Industrial Economics*, March 1995, 43(1), 13–48.

Da Silveira, G. Innovation diffusion: Research agenda for developing economies. *Technovation*, December 2001, 21(12), 767–773.

D'Aspremont, C. and Jacqueminde, A. Cooperative and non-cooperative R&D in a Duopoly with spillover. *American Economic Review*, December 1988, 78(5), 1133–1137.

Dang, Y., Liu, S., and Liu, B. Study on the multi-attribute decision model of grey target based on interval number. *Engineering Science*, 2005, 7(8), 31–35.

Dang, Y., Liu, S., Liu, B., and Zhai, Z. Study on the integrated grey clustering method under the clustering coefficient with non-distinguished difference. *Chinese Journal of Management Science*, 2005, 13(4), 69–73.

Dang, Y., Liu, S., and Mi, C. Multi-attribute grey incidence decision model for interval number. *Kybernetes: The International Journal of Systems & Cybernetics*, 2006, 35(7/8), 1265–1272.

De Bresson, C. Predicting the most likely diffusion sequence of a new technology through the economy: The case of superconductivity. *Research Policy*, September 1995, 24(5), 685–705.

Decanio, S.J. and Laitner, J.A. Modeling technological change in energy demand forecasting: A generalized approach. *Technological Forecasting and Social Change*, July 1997, 55(3), 249–263.

Deng, J. *Gray System Basic Methods*. Huazhong University of Science and Technology Press, Wuhan, China, 1987.

Du, C. and Yang, L. A research for the concept of technical innovation, technical progress and technical spread. *China Economist*, 2002, 3, 43–44.

Duan, L. and Liu, S. A study on the spill effect model in the diffusion field. *Journal of Nanjing University of Aeronautics and Astronautics (Social Sciences)*, 2003, 5(1), 32–36.

Duan, L. and Liu, S. Theoretical study on the model of technical diffusion state in technical diffusion field. *Journal of Beijing Polytechnic University*, 2003, 29(2), 251–256.

Duan, L. and Liu, S. Theoretical studying of the technology diffusion speed model in technology diffusion field. *Journal of Northwest SCI-TECH University of Agriculture and Forestry*, 2003, 3(3), 45–48.

Duan, L.-z. and Liu, S.-f. Theoretical study on the model of technical diffusion state in technical diffusion field. *Journal of Beijing Polytechnic University*, 2003, 5(1), 32–36.

Duan, L.-z. and Liu, S.-f. Theoretical studying of the technology diffusion speed model in technology diffusion field. *Journal of Northwest SCI-TECH University of Agriculture and Forestry*, 2003, 5(1), 32–36.

Dunning, J.H. and Gugler, P. *Technology-Based Cross-Border Alliances*. The Globalization of Business, Routledge, London, U.K., 1993, pp. 191–219.

Duysters, G. and Hagedoorn, J. Internationalization of corporate technology through strategic partnering: An empirical investigation. *Research Policy*, January 1996, 25(1), 1–12.

Fan, Z. Brief analysis of technical spread and overflow effect of transnational corporation. *Science of Science and Management of S & T*, 2003, 4, 52–54.

Fan, X., Chen, H., and Yang, S. Analysis on the meaning of technology transfer and its relative concept. *Science and Technology Management Research*, 2000, 6, 44–46.

Fang, Z., Liu, S., and Mi, C. Study on optimum value of grey matrix game based on token of grey interval number, in *The 6th World Congress on Intelligent Control and Automation*.

Fang, Z., Liu, S., and Shi, H. The calculating model and positivist analysis of productivity and its fluctuating period leading GDP's, in *The 5th International Institute for General Systems Studies*.

Fang, Z., Liu, S., Wu, X., Zhang, H., and Mi, C. Study on matrix solution method of grey matrix game based on full rank grey payoff matrix, in *2006 IEEE Conference on Systems, Man, and Cybernetics*.

Fang, Z., Liu, S., Ruan, A., and Zhang, X. Study on venture problem of potential optimal pure strategy solution for grey interval number matrix game. *Kybernetes: The International Journal of Systems & Cybernetics*, 2006, 35(7/8), 1273–1283.

Fisher-Vanden, K. Management structure and technology diffusion in Chinese state-owned enterprises. *Energy Policy*, February 2003, 31(3), 247–257.

Fruin, W.M. and Prime, P. Competing strategies of FDI and technology transfer to China: American and Japanese firms. The William Davidson Institute at the University of Michigan Business School, Working Paper Number 218, January 1999.

Fu, Q. System analysis & model of technology transfer. Chongqing University, 1999.

Gerchak, Y. and Barlar, M. Allocation resources to research and development project in a competitive environment. *IIE Transaction*, 1999, 31(9), 827–834.

Geroski, P.A. Models of technology diffusion. *Research Policy*, April 2000, 29(4–5), 603–625.

Gong, G. and Keller, W. Convergence and polarization in global income levels: A review of recent results on the role of international technology diffusion. *Research Policy*, June 2003, 32(6), 1055–1079.

Gong, Z. and Liu, S. Consistency and priority of triangular fuzzy number complementary judgment matrix. *Control and Decision*, 2006, 21(8), 903–907.

Gong, Y., Cai, S., and Zhang. J. Influence on enterprise management strategy of game theory. *China Soft Science*, 1999, 8, 74–76.

Grether, J.-M. Determinants of technological diffusion in Mexican manufacturing: A plant-level analysis world development. July 1999, 27(7), 1287–1298.

Gritsevskyi, A. and Nakicenovi, N. Modeling uncertainty of induced technological change. *Energy Policy*, November 2000, 28(13), 907–921.

Guan, J. Discuss the study on high-tech diffusion model once more. *Science of Science and Management of S & T*, 1995, 16(8), 30–36.

Guan, J. and Zhang, X. On the concept of innovation diffusion and its research methods. *Chemical Techno Economics*, 1996, 1, 23–27.

Guice, J. Designing the future: The culture of new trends in science and technology. *Research Policy*, January 1999, 28(1), 81–98.

Gumbau-Albert, M. Efficiency and technical progress: Sources of convergence in the Spanish regions. *Applied Economics*, 2000, 32(4), 479–489.

Guo, S., Liu, S., and Fang, Z. Game analysis of the best technological contents of oligopsony enterprise's technological innovation. *Science & Technology Progress and Policy*, 2007, 24(7), 111–114.

Guo, X., Liu, S., and Fang, Z. Influence on research and invention activities in manufacturing in our country by foreign technology pervasion. *Science & Technology Progress and Policy*, 2008, 25(5), 69–71.

Hagedoorn, J. Strategic technology partnering in the 1980s: Trends, networks, and corporate patterns in non-core technologies. *Research Policy*, March 1995, 24(2), 207–231.

Hahn, Y.-H. and Yu, P.-I. Towards a new technology policy: The integration of generation and diffusion. *Technovation*, January 1999, 19(3), 177–186.

Harabi, N. Channels of R&D spillovers: An empirical investigation of Swiss firms, *Technovation*, November 12, 1997, 17(11–12), 627–635, 724–725.

Hattori, H. and Shishido, T. Spillover of hydrogen over zirconium oxide promoted by sulfate ion and platinum. *Applied Catalysis A: General*, October 22, 1996, 146(1), 157–164.

Hauknes, J. Innovation in the service economy. STEP Report, 1996, 7, STEP Group, Oslo, Norway.

He, B., Gu, J., and Yan, Y. *Technology Transfer & Technology Advancement in China*. Economic Management Publishing House, Beijing, China, 1996.

Hollifield, C.A. and Donnermeyer, J.F. Creating demand: Influencing information technology diffusion in rural communities. *Government Information Quarterly*, May 2003, 20(2), 135–150.

Hu, R. and Wang, Q. Epidemic model and application of technology diffusion. *Journal of Agrotechnical Economics*, 1996, 6, 52–53.

Hu, Z. and Yie, C. Infectious disease model in new technology dissemination and its empirical evidence. *Journal of Wuhan University of Technology*, 1998, 20(2), 76–78.

Hu, Y., Fang, Z., Liu, S., and Lu, F. Study on model of evolutionary game chain of industry cluster based on asymmetry case, in *2005 IEEE International Conference on Industrial Technology*.

Huang, J. The survival way and technological of innovation private enterprise. *Journal of South China University of Technology (Social Science)*, February 2005, l7(1), 28–33.

Hur, K.I. and Watanabe, C. Unintentional technology spillover between two sectors: Kinetic approach. *Technovation*, April 2001, 21(4), 227–235.

Jacobsson, S. and Johnson, A. The diffusion of renewable energy technology: An analytical framework and key issues for research. *Energy Policy*, July 31, 2000, 28(9), 625–640.

Jha, R. and Majumdar, S.K. A matter of connections: OECD telecommunications sector productivity and the role of cellular technology diffusion. *Information Economics and Policy*, September 1999, 11(3), 243–269.

Jia, J. and Shao, X. Discuss technology transfer guided by market. *Soft Science*, 1999, (s1), 47–49.

Jian, L. and Liu, S. A hybrid approach of VPRS and PNN to knowledge discovery. *Journal of the China Society for Scientific and Technical Information*, 2005, 24(4), 426–432.

Jianguo, L. The present situation and question of our country technology transfer. *China Economic and Trade Herald*, July 1997, 13, 22–23.

Johansen, S. Statistical analysis of co-integration vectors. *Journal of Economic Dynamics and Controls*, March 1988, 12(2–3), 231–254.

Johansen, S. and Juselius, K. Maximum likelihood estimation and inference on co-intergration—With applications to the demand for money. *Oxford Bulletin of Economics and Statistics*, February 1990, 52(2), 169–210.

Jun, D.B. and Park, Y.S. A choice-based diffusion model for multiple generations of products. *Technological Forecasting and Social Change*, May 1999, 61(1), 45–58.

Kang, R. The new pattern of China technology pattern in the 90s. *Management World*, January 1994, 1, 169–171.

Kathuria, V. Technology transfer for GHG reduction: A framework with application to India. *Technological Forecasting and Social Change*, May 2002, 69(4), 405–430.

Kiiski, S. and Pohjola, M. Cross-country diffusion of the Internet. *Information Economics and Policy*, June 2002, 14(2), 297–310.

Kim, N. and Srivastava, R.K. Managing intraorganizational diffusion of technological innovations. *Industrial Marketing Management*, May 1998, 27(3), 229–246.

Klibanoff, P. and Morduch, J. Decentralization, externalities, and efficiency. *Review of Economic Studies*, April 1995, 62(2), 223–247.

Kokko, A. Technology, market characteristics, and spillovers. *Journal of Development Economics*, 1994, 43, 279–293.

Kotaro, S. Cooperative and non-cooperative R&D in a oligopoly with spillovers. 1992, (5).

Kurt, H. and Norman, G. Managing International Technology Transfer, pp. 221–224.

Kwasnicki, W. and Kwasnicka, H. Long-term diffusion factors of technological development: An evolutionary model and case study, *Technological Forecasting and Social Change*, May 1996, 52(1), 31–57.

Laranja, M. and Fontes, M. Creative adaptation: The role of new technology based firms in Portugal. *Research Policy*, April 1998, 26(9), 1023–1036.

Lee, J. Small firms innovation in two technology settings. *Research Policy*, May 1995, 24(3), 391–401.

Li, P. and Wang, W. Analysis of spillover effect in duality technology. *Nankai Economic Studies*, 1998, 5, 33–38.

Li, P. Analysis of spillover effect in technology diffusion. *Nankai Journal*, 1999, 2, 28–32.

Li, P. *Technology Diffusion Theory and Empirical Research*. Shanxi Economic Press, Xi' an,1999.

Li, S. Quantitative analysis of elements affect regional science & technology resource collocation efficiency in China. *Scientific Management Research*, 2003, 21(2), 61–62.

Li, B. and Liu, S. Evaluation on time series grey clustering of technology comprehensive strength in Jiang Su Province. *Industrial Technology & Economy*, 2004, 23(2), 59–61.

Li, B. and Liu, S. The hierarchic grey incidence analysis for the science & technology system in the city (territory) hierarchy of Jiangsu Province. *Science & Technology Progress and Policy*, 2005, 22(2), 63–65.

Li, Q. and Liu, S. The foundation of the grey matrix and the grey input-output analysis. *Applied Mathematic Modeling*, 2008, 32, 267–291.

Li, S. and Liu, S. Analysis of high-tech industry R&D cost in China. *Statistics and Decision*, 2007, 4, 64–66.

Li, J. and Zhou, S. Optimization research of matching scale between technology human resource and financial resource in China. *Scientific Management Research*, 2001, 6, 72–76.

Li, Y., Shi, P., and Liu, Y. *The Project Supported by NSFC*. Shanxi People's Publishing House, Xi' an, 2001.

Li, Q., Liu, S., and Dang, Y. Extension analysis to adjust the industrial structure of rural economy in China. *Journal of Harbin Institute of Technology*, 2006, 38(7), 1205–1208.

Liang, K. and Li, L.S. Problems and countermeasures in adopting tax policy to accelerate market acceptance of scientific and technological achievements in China. *Journal of Southeast University Philosophy and Social Science Edition*, November 2005, 7(6), 27–31.

Liu, J. On the dynamics of stochastic diffusion of manufacturing technology *European Journal of Operational Research*, August 2000, 124(3), 601–614.

Lin, J. Trilogy of high efficacy team: Talent flow, work flow, knowledge flow. *IT World Newspaper*, 2004, 9, 18–23.

Liu, S. On index system and mathematical model for evaluation of scientific and technical strength. *Kybernetes: The International Journal of Systems & Cybernetics*, 2006, 35(7/8), 1256–1264.

Liu, S. and Chen, K. Analysis on distribution structure and utilization efficiency of the funds for science and technology in China. *Chinese Journal of Management Science*, 2002, 10, 40–43.

Liu, B. and Cheng, X. Spillover effect analysis of technology introduction. *Coal Economic Research*, 1999, 5, 4–7.

Liu, S. and Dang, Y. Analysis on distribution structure and employ efficiency of the funds for science and technology in Henan Province. *Science of Science and Management of S & T*, 2001, 22(10), 12–15.

Liu, S. and Dang, Y. Technical change and the funds for science and technology. *Kybernetes: The International Journal of Systems & Cybernetics*, Beijing, 2004, 33(2), 295–302.

Liu, L. and Hu, S. Comparative study on R&D Management at home and broad & its enlightenment for technology investment allocation of China. *Studies in Science of Science*, 2000, 18(1), 62–66.

Liu, R. and Hu, X. Dialectical relationship preliminary of modern science and technology innovation and diffusion. *Science & Technology Progress and Policy*, 2001, 4, 110–113.

Liu, S. and Lin, Y. On measures of information content of grey numbers. *Kybernetes: The International Journal of Systems & Cybernetics*, 2006, 35(6), 899–904.

Liu, J.-q. and Liu, Z.-g. Regimes switching in China's business cycle and analysis of the properties of regimes. *Journal of Zhejiang University (Humanities and Social Sciences)*, 2006, 3, 95–102.

Liu, S. and Wang, R. Mechanism, effect and countermeasure of technology talent agglomeration. *Journal of Nanjing University of Aeronautics & Astronautics (Social Sciences)*, 2008, 10(1), 47–51.

Liu, X., Wang, Y., and Zhang, H. On the way of technical expansion and spill over for transnational companies. *Journal of Dongbei University of Finance and Economics*, 2003, 25(1), 28–30.

Liu, S., Dang, Y., and Lin, Y. Synthetic utility index method and venturous capital decision-making. *Kybernetes: The International Journal of Systems & Cybernetics*, 2004, 33(2), 288–294.

Liu, S., Dang, Y., and Fang, Z. *Grey Information: Theory and Practical Applications* (Third Annual Edition). Science Press, 2004(11).

Liu, S., Li, B., and Dang, Y. The G-C-D model and technical advance. *Kybernetes: The International Journal of Systems & Cybernetics*, 2004, 33(2), 303–309.

Liu, S., Tang, X., Yuan, C., and Dang, Y. Study on order degree of industrial structure in China. *Economics Dynamics*, 2004, 5, 53–56.

Liu, S., Shi, H., and Fang, Z. The pareto optimal distribution of linear programming model of technology flow of uni-polar city based on the perfect mechanism, in *2007 IEEE International Conference on Grey System and Intelligent Services*, 2008.

Lou, C. Innovative technology information transmit mechanism of technology innovation and diffusion. *Science & Technology Progress and Policy*, 1999, 16(5), 50–52.

Lu, F., Liu, S., Fang, Z., and Hu, Y. Study on chain structure model of industry clusters grey evolutionary game based on symmetric case, in *2006 IEEE Conference on Systems, Man, and Cybernetics*.

Luo, D. and Liu, S.-f. Research on grey multi-criteria risk decision-making method. *Systems Engineering and Electronics*, 2004, 26(8), 1057–1059.

Luo, D. and Liu, S. A decision-making method for grey portfolio selection model. *Mathematics in Practice and Theory*, 2005, 35(6), 37–43.

Mansfield, E. *Technological Change: An Introduction to a Vital Area of Modern Economics*. Norton Publishing House, New York, 1971.

Masini, A. and Frankl, P. Forecasting the diffusion of photovoltaic systems in southern Europe: A learning curve approach. *Technological Forecasting and Social Change*, January 2003, 70(1), 39–65.

Mazzoleni, R. Learning and path-dependence in the diffusion of innovations: Comparative evidence on numerically controlled machine tools. *Research Policy*, December 1997, 26(4–5), 405–428.

McKendrick, D.M. Source of imitation: Improving bank process capabilities. *Research Policy*, September 1995, 24(5), 783–802.

Mei, J.J. and Li, S. Mathematical model of optimizing stage structure of science & technology resources allocation. *Journal of Beijing Institute of Machinery*, 2002, 17(2), 61–63.

Meng, Y. *Economics Social Field Theory*. Renmin University of China Press, Beijing, 1999.

Mi, C., Liu, S., and Yang, J. Research on the relationship between Jiangsu province's S&T investment and economic growth: Based on the relative degree of grey incidences. *Science of Science and Management of S & T*, 2004, 25(1), 34–36.

Miles, I., Kastrinos, N., and Bilderbeek, R.P. Knowledge-intensive business services-users, carriers and sources of innovation. EIMS publication No. 15: EC, 1995.

Miller, D. and Garnsey, E. Entrepreneurs and technology diffusion: How diffusion research can benefit from a greater understanding of entrepreneurship. *Technology in Society*, November 2000, 22(4), 445–465.

Morgenstern, R.D. and Al-Jurf, S. Can free information really accelerate technology diffusion? *Technological Forecasting and Social Change*, May 1999, 61(1), 13–24.

Nakamura, T. International knowledge spillovers and technology imports: Evidence from Japanese chemical and electric equipment industries. *Journal of the Japanese and International Economies*, September 2001, 15(3), 271–297.

Neij, L. Use of experience curves to analyze the prospects for diffusion and adoption of renewable energy technology. *Fuel and Energy*, January, 1998, 39(1), 42 (abstracts).

Nilakanta, S. and Scamell, R.W. The effect of information sources and communication channels on the diffusion of innovation in a data base development environment. *Management Science*, January 1990, 36(1), 24–40.

Ning, X. and Liu, S. *Management Prediction and Decision Method*. Science and Technology Publishing House, Beijing, 2003.

Okejiri, E. Foreign technology and development of indigenous technological capabilities in the Nigerian manufacturing industry. *Technology in Society*, April 2000, 22(2), 189–199.

Papaconstantinou, G., Sakurai, N., and Wyckoff, A. Domestic and international product-embodied R & D diffusion. *Research Policy*, July 1998, 27(3), 301–314.

Parente, S.L. and Prescott, E.C. Barriers to technology adoption and development. *Journal of Political Economy*, April 1994, 102(2), 298–321.

Park, Y.-T. Technology diffusion policy: A review and classification of policy practices. *Technology in Society*, August 1999, 21(3), 275–286.

Pavlinek, P. Regional development implications of foreign direct investment in central Europe. *European Urban and Regional Studies*, 2004, 11(1), 47–70.

Podobnik, B. Toward a sustainable energy regime: A long-wave interpretation of global energy shifts. *Technological Forecasting and Social Change*, November 1999, 62(3), 155–172.

Poyago Theotoky, J. A equilibrium and optimal size of a research oint venture in an oligopoly with spillovers. *Journal of Industrial Economics*, June 1995, 43(2), 20–29.

Reinhardt, N. Latin America's new economic model: Micro responses and economic restructuring. *World Development*, September 2000, 28(9), 1543–1566.

Ren, Y., Liu, S., and Fang, Z. Building theory on productive forces of science and technology in the times of knowledge economy. *Productivity Research*, 2006, 8, 76–78.

Ren, Y., Liu, S., and Fang, Z. Research on science and technology productivity and construction of conceptual architecture related to it. *Science & Technology Progress and Policy*, 2006, 8, 120–122.

Rennings, K. Redefining innovation—Eco-innovation research and the contribution from ecological economics. *Ecological Economics*, February 2000, 32(2), 319–332.

Reppelin-Hill, V. Trade and environment: An empirical analysis of the technology effect in the steel industry. *Journal of Environmental Economics and Management*, November 1999, 38(3), 283–301.

Rogers, E. *Diffusion of Innovations*. The Free Press, New York, 1995.

Rong, L., Huang, B., and Hu, J. Urban and rural social economy resource allocation & overall planning mechanism and policy of technology elements flow in Shandong province. *Journal of Shandong Agricultural University* (*Social Science Edition*), 2006, 4, 18–23.

Rosana, G. and C. Roger. The role knowledge in resource allocation to exploration verse exploitation in technologically oriented organization. *Decision Science*, 2003, 34(2), 323–349.

Ruan, A., Liu, S., and Fang, Z. Study on development kinetic energy of Chinese industry and measure model of energy, in *6th World Congress on Intelligent Control and Automation*. June 2006, Daliau, China.

Ruan, A. and Liu, S. Industrial cluster growth research based on evolutionary game model. *Science of Science and Management of S & T*, 2008, 2, 91–95.

Ruan, A., Liu, S., and Fang, Z. Category of symmetric grey evolutionary game under strong correlation between individual pay-off and evolution stage. *Control and Decision*.

Sheng, Y. Theory and practical problems of technology transfer in China. *Scientific Management Research*, 1994, 12(6), 40–43.

Shi, H., Liu, S., and Fang, Z. Study on the model of grey matrix game based on grey mixed strategy—Properties of grey mixed strategy and model of linear program, in *The 7th World Congress on Intelligent Control and Automation*.

Shi, H., Liu, S., Du, H., and Xu, X. Measurement and analysis of contribution of economic factors to economic growth in six provinces and Shanghai City of East China based on the same technique, in *2006 IEEE Conference on Systems, Man, and Cybernetics*, 2006.

Shi, H., Liu, S., and Fang, Z. Study on transfer calculation in manufacturing of electron and communication equipment based on gravitation model. *Industrial Technology & Economy*, 2007, 26(8), 104–107.

Shi, H., Liu, S., and Fang, Z. Research on flowing mechanism of the technical productive forces. *Science of Science and Management of S & T*, 2007, 28(11), 25–28.

Shi, H., Liu, S., Fang, Z., Cheng, Y., and Zhang, H. The model of grey periodic incidence and their rehabilitation, in *2007 IEEE International Conference on Systems, Man and Cybernetics*, 2007.

Shi, H., Liu, S., Fang, Z., and Ruan, A. Linear programming model with uni-polar city for different levels allocation of productivity of science and technology based on the perfect mechanism, in *The 7th World Congress on Intelligent Control and Automation*, 2008.

Shi, H., Liu, S., and Fang, Z. The calculating model and empirical analysis of science & technology productivity and its leading GDP's fluctuating cycle, in *2007 IEEE International Conference on Grey System and Intelligent Services*, 2008.

Shoham, Y. Agent-oriented programming. Artificial intelligent. 1993, 60, 51–92.

Slade, E.P. and Anderson, G.F. The relationship between per capita income and diffusion of medical technologies. *Health Policy*, October 2001, 58(1), 1–14.

Sohn, S.Y. and Ahn, B.J. Multigeneration diffusion model for economic assessment of new technology. *Technological Forecasting and Social Change*, March 2003, 70(3), 251–264.

Steensma, H.K. and Fairbank, J.F. Internalizing external technology: A model of governance mode choice and an empirical assessment. *The Journal of High Technology Management Research*, Spring 1999, 10(1), 1–35.

Stock, G.N. and Tatikonda, M.V. A typology of project-level technology transfer processes. *Journal of Operations Management*, November 2000, 18(6), 719–737.

Stockmann, R. The sustainability of development projects: An impact assessment of German vocational-training projects in Latin America. *World Development*, November 1997, 25(11), 1767–1784.

Su, J. and Wang, Y. Design & management of investment mechanism in industrialization of high-tech. *Journal of Dalian University of Technology (Social Sciences)*, 1999, 79(20(1)), 36–38.

Suzumura, K. Cooperative and noncooperative R&D in an oligopoly with Spillovers. *American Economic Review*, December 1992, 82(5), 1307–1320.

Tang, Y. The economic problem must study in the technology Introduction. *World Economy*, January 1978, 1, 69–71.

Terttu, L. and Bertel, S. Quality evaluation in management of basic and applied research. *Research Policy*, 1990, 19, 357–368.

Todo, Y. Empirically consistent scale effects: An endogenous growth model with technology transfer to developing countries. *Journal of Macroeconomics*, March 2003, 25(1), 25–46.

Traxler, G. and Byerlee, D. Linking technical change to research effort: An examination of aggregation and spillovers effects. *Agricultural Economics*, March 2001, 24(3), 235–246.

Van Elkan, R. Catching up and slowing down: Learning and growth parerna in an open economy. *Journal of International Economics*, August 1996, 41(1–2), 95–111.

Vanotras, N.S. The challenging economic context: Strategic alliances among multinationals, *Center for Science and Technology Policy*, Rensselaer Polytechnic Institute, 1989.

Verspagen, B. and DeLoo, I. Technology spillovers between sectors. *Technological Forecasting and Social Change*, March 1999, 60(3), 215–235.

Vessuri, H. Science, politics, and democratic participation in policy-making: A Latin American view. *Technology in Society*, 2003, 25, 263–273.

Vonotras, N.S. Inter-firm cooperation in imperfectly appropriable research: Industry performance and welfare implications. New York University, New York, 1989.

Walker, P.J., Calalano, D.R.F. et al. Diversity and power in the world city network. *Cities*, 2002, 19(4), 231–241.

Wang, J. *Evaluate Theory and Practise of Technology Advancement*. Science and Technology Literature Publishing House, 1986.

Wang, C. Three characteristics of technology diffusion—Transnational enterprises and international technology transfer. 2000, 16, 18–19.

Wang, M. and Chen, Y. Mechanics of increasing return and implication of development strategy in the information industry. *Journal of Renmin University of China*, 1999, 4, 13–18.

Wang, Y. and Jiao, B. Foreign capital utilization mode & technology diffusion. *Journal of Shanxi University of Finance and Economics*, 1998, 3, 21–24.

Wang, X. and Li, P. Spillover effect in theory model. *Nankal Economic Studies*, 1995, 4, 61–64.

Wang, Y. and Liu, S. Construction of new financial support system, advancement of technology productivity. *Finance and Economics*, 2005, 3, 55–56.

Wang, R. and Liu, S. Drive mechanism of innovative talents in developed regions. *Jiangsu Rural Economy*, 2006, 3, 49–50.

Wang, P. and Liu, S. Study on advantage and capacity of self-innovation in high-tech zones base on industrial cluster. *Enterprise Economy*, 2008, 5, 44–46.

Wang, Y. and Liu, S. Empirical study on international technology spillover channels. *The Journal of Quantitative & Technical Economics*, 2008, 4, 153–161.

Wang, Y. and Liu, S. The influence of OFDI on Chinese industrial structure: An analysis based on the grey incidence theory. *World Economy Study*, 2008, 4, 61–66.

Wang, M., He, P., Zhu, Q., and Ding, X. The economic effect and policy mechanism of the essential factors' flowing in the east–west cooperation. *China Soft Science*, 2001, 7, 104–107.

Wang, R., Liu, S., and Dang, Y. Study on indices system of regional development goals for science and technology. *Forum on Science and Technology in China*, 2005, 6, 79–82.

Wang, R., Liu, S., and Gu, J. Cost and benefit analysis of innovative talents flow in Jiangsu province. *Industrial Technology & Economy*, 2006, 25(3), 90–91, 101.

Watanabe, C., Zhu, B., Griffy-Brown, C., and Asgari, B. Global technology spillover and its impact on industry's R&D strategies. *Technovation*, May 2001, 21(5), 281–291.

Wei, Y.-m., Zeng, R., Fan, Y., Tsai, H., Xu, W.-x., and Fu, X.-f. A multi-objective goal programming model for Beijing's coordination development of population, resources, environment and economy. *Systems Engineering-Theory & Practice*, 2002, 2(2), 74–83.

Weiss, J.A. and Dale, B.C. Diffusing against mature technology: Issues and strategy. *Industrial Marketing Management*, July 1998, 27(4), 293–304.

Wooldridge, M. Agent-based software engineering. *IEEE Proceedings—Software Engineering*, February 1997, 144(1).

Wu, H. and Liu, S. Evaluation on the R&D relative efficiency of different areas in China based on improved DEA model. *R & D Management*, 2007, 19(2), 108–112, 128.

Wu, C., Dai, D., and Sujingqin. *Technology Innovation Diffusion*. Chemical Industry Press, Beijing, China, 1997.

Wu, H. and Zheng, C. An empirical analysis on the relative efficiency of the input-output of science and technology. *Scientific Management Research*, 2003, 21(3), 93–96.

Wu, J., Liu, S., and Shi, Q. Study on new knowledge innovative system and knowledge transfer based on industrial cluster. *Enterprise Economy*, 2007, 3, 33–35.

Wu, J., Liu, S.-f., and Shi, Q.-f. Research on knowledge innovation and transformation mechanism of shipping enterprises based on entropy theory. *Ship Engineering*, 2007, 29(1), 34–37.

Wu, L. Studies on the world technology zone innovation pattern compare. *Forum on Science and Technology in China*, 2002, 1, 36–40.

Xie, F. Theory, model, demonstration of foreign direct investment and technology advancement in China. Fudan University, 1999.

Xie, H., He, J., and Zheng, C. Study of technical diffusion model on the theory of technical environment. *Forecasting*, 1999, 5, 57–59.

Xie, H., He, J., and Zheng, C. Technology diffusion theory on the basis of technical environment. *Forecasting*, 1999, 5, 57–69.

Xu, B. Multinational enterprises, technology diffusion, and host country productivity growth. *Journal of Development Economics*, August 2000, 62(2), 477–493.

Xu, Q. and Sheng, Y. Research summary of technology transfer at home and broad. *Scientific Management Research*, 1993, 11(4), 11–14.

Xu, X., Fang, Z., Ma, S., and Zhang, X. The measurement and calculation of technology progresses contribution rate of Chinese provinces basing on the same technology base-point, in *2006 IEEE Conference on Systems, Man, and Cybernetics*.

Yao, T. and Liu, S. Evaluation of technology research in university based on gray combination entropy weight. *Journal Of Information*, 2008, 7, 93–94, 99.

Yi, X. Element fluxion and region unequilibrium development. *Journal of Hebei University (Philosophy and Social Science)*, 2004, 5, 94–96.

Yin, X. Manufacturing: "Multiplier effect" spillover and technology upgradation. *Economist*, 1998, 5, 101–108.

Yuan, G. Local economical difference, macro economic fluctuation and macroscopic policy effect. *Urban Finance Forum*, June 1997, 6, 2–7.

Yue, H. and Liang, L. Comparison concerning regional distribution of R&D input between China and USA. *Science Research Management*, 2001, 22(6), 28–35.

Yue, H., Liu, S., and Liang, L. A bibliometric analysis on the technology innovation research in China. *Science Research Management*, 2008, 29(3), 43–52.

Zhang, W. *Game Theory & Information Economics*. Shanghai Joint Publishing, Shanghai People's Publishing House, Shanghai, 1996.

Zhang, Y., Li, Z., and Liu, L. Diffusion of agricultural scientific and technological innovation, factor flow and scale economy in China. *Journal of Industrial Engineering and Engineering Management*, 2001, 15(3), 24–28.

Zhao, L. *Theory of Technology Transfer*. Science and Technology of China Press, Beijing, 1992.

Zhao, R. Relative value relevance of R&D reporting: An international comparison. *Journal of International Financial Management and Accounting*, 2002, 13(2), 153–174.

Zhou, T. *Advanced Development Economics*. China Renmin University Press, Beijing, 2006.

Zhou, Z. and Liu, S. Research on valuation of venture capital with grey prediction. *Journal of Nanjing University of Aeronautics & Astronautics*, 2004, 36(5), 644–648.

Zhou, H. and Liu, S. The knowledge management strategy of enterprise in new economic environment. *Economic Problems*, 2005, 1, 37–39.

Zhou, H. and Liu, S.-f. Endogenous character and dynamic character sources of sustainable corporate competitive advantage. *Forecasting*, 2006, 25(3), 38–42.

Zhao, L., Leng, X., and Duan, L. *City Innovation System*. Tianjin University Press, Tianjin 2002.

Zhu, J. and Fu, M. Probing and analysis of model for technology transfer. *Journal of Beijing University of Aeronautics and Astronautics*, 1997, 23(2), 229–235.

Ziss, S. Strategic R&D with spillovers, collusion and welfare, 1994, (4).

Ziss, S. Strategic R&D with spillovers, collusion and welfare. *The Journal of Industrial Economics*, December 1994, 42(4), 375–393.

Ziss, S. Strategic R&D with spillovers, collusion and welfare. *The Journal of Industrial Economics*, December 1994, 42(4), 55–63.

Zou, X. Analysis of flowing mechanism of capital factors. *Journal of Chongqing Technology and Business University*, 2004, 21(4), 19–22.

Index

D

E

F

G

N

P

Q

R

S

T

U

V